Colonial Technology

Studies in Australian History

Series editors:
Alan Gilbert, Patricia Grimshaw and Peter Spearritt

Steven Nicholas (ed.)   *Convict Workers*
Pamela Statham (ed.)   *The Origins of Australia's Capital Cities*
Jeffrey Grey   *A Military History of Australia*
Alastair Davidson   *The Invisible State*
James A. Gillespie   *The Price of Health*
David Neal   *The Rule of Law in a Penal Colony*
Sharon Morgan   *Land Settlement in Early Tasmania*
Audrey Oldfield   *Woman Suffrage in Australia*
Paula J. Byrne   *Criminal Law and Colonial Subject*
Peggy Brock   *Outback Ghettos*
Raelene Frances   *The Politics of Work*
Luke Trainor   *British Imperialism and Australian Nationalism*
Margaret Maynard   *Fashioned from Penury*
Dawn May   *Aboriginal Labour and the Cattle Industry*
Joy Damousi and Marilyn Lake (eds)   *Gender and War*
Michael Roe   *Australia, Britain, and Migration, 1915–1940*
John Williams   *The Quarantined Culture*
Nicholas Brown   *Governing Prosperity*

Colonial Technology

Science and the Transfer of Innovation to Australia

Jan Todd

CAMBRIDGE
UNIVERSITY PRESS

CAMBRIDGE UNIVERSITY PRESS
Cambridge, New York, Melbourne, Madrid, Cape Town, Singapore, São Paulo, Delhi

Cambridge University Press
The Edinburgh Building, Cambridge CB2 8RU, UK

Published in the United States of America by Cambridge University Press, New York

www.cambridge.org
Information on this title: www.cambridge.org/9780521109840

First published 1995
This digitally printed version 2009

A catalogue record for this publication is available from the British Library

National Library of Australia cataloguing-in-publication data

Todd, Jan (Janene), 1947–.
Colonial technology: science and the transfer of
innovation to Australia.
Bibliography.
Includes index.
1. Technology – Australia – History. 2. Technology
transfer – Australia – History. 3. Science – Australia –
History. I. Title. (Series: Studies in Australian
history (Cambridge, England)).
609.94

Library of Congress Cataloguing in Publication data

Todd, Jan, 1947–.
Colonial technology: science and the transfer of innovation to
Australia/Jan Todd.
    p.   cm. – (Studies in Australian history)
Includes bibliographical references and index.
1. Technology transfer – Australia. 2. Technology and state –
Australia. 3 Science and state – Australia. I. Title.
II. Series.
T174.3.T644 1995
338.9'9406–dc20   95–6114

ISBN 978-0-521-46138-2 hardback
ISBN 978-0-521-10984-0 paperback

Contents

Figures and Tables

Currency and Measurement

To retain historical accuracy the text contains the units of currency and measurement as used at the time.

Reference to sums of money are in pounds (£). There were 12 pennies (d) in one shilling (s) and 20 shillings in one pound. It is difficult to compare the value of money in the nineteenth century with today's prices. Inflation has caused such changes that, for example, household goods worth 5 cents in 1900 would have cost $1 in 1980, and about $2 in 1990.

Measurements in the text are mainly given in imperial measures, though metric units are given when these were used in the historical sources. Conversions are as follows.

Length
1 inch = 25.4 mm
1 foot = 30.5 cm
1 mile = 1.61 km

Mass
1 grain (gr) = 0.065 g
1 pennyweight (dwt) = 1.4 g
1 ounce (oz) = 28.3 g
1 pound (lb) = 454 g
1 hundredweight (cwt) = 50.8 kg
1 ton = 1.02 t

Area
1 acre = 0.405 ha

Volume
1 pint = 568 mL

Temperature
To convert degrees Fahrenheit to degrees centigrade, subtract 32, multiply by 5, and divide the result by 9.

Preface

In the late twentieth century, Australia struggles with the challenges of the international economy: attracted by the market opportunities, daunted by the competition, confused about just how much to open the economy and how much to level the 'playing field'.

One of the most critical problems is just how to integrate a hitherto protected domestic production into the international economy as globalisation proceeds apace. Many contemporary commentators suggest that a range of vital domestic linkages are missing: linkages from our centres of scientific and technical research to users who can convert it into economically viable and internationally competitive technology; linkages among interrelated industries which would benefit from the synergy; linkages from public-sector technical activities to private-sector commercialisation; linkages from domestic production to international marketing; linkages from industrial needs for expertise to responsive training institutions.

Some say the reasons for the current malaise lie in the way Australia became integrated into the international economy in the nineteenth century as a dependent and marginalised colonial offshoot of Britain. The allotted role of raw-material supplier to advanced industrial nations implied many 'missing links' in the domestic economy, a situation which remains entrenched today, when natural resource commodities return declining rewards.

This book began as an exploration of that proposition. The result reveals some important links which were forming in Australia in the late nineteenth century as it was drawn into the expanding international economy to an extent which belied its small size. A most vital connection was from imported technology to the domestic technology and economy, and two of the most important routes were through the link from local science to local productive technology and through the link from public sector to private.

It is hoped that two types of readers will find this book of interest and relevance: those who like a good story about Australia's history, and

x *Contents*

those who are looking for lessons of wider significance from the history
of science and technology. The story comes in the form of two interest-
ing sagas of technological and industrial development in Parts II and III.
The first, about microbes, rabbits and sheep, traces the transfer of
anthrax vaccination into the Australian pastoral industry; the second, on
rocks, cyanide and gold, follows the trail of the cyanide process of gold
extraction into Australian mining. In order to draw out the wider implications, Part I will set the scene and the questions, while Part IV will
explore the answers.

Some of my work for this book has already appeared as journal articles:
in particular, some of the material presented in Chapter 5 was published
in *Prometheus* (June 1989) and some of the material in Chapter 14 was
published in *Annals of Science* (volume 50, 1993). I thank the publishers
of those journals for their permission to present the material again here.

Four people stand out for their assistance in the preparation of this
book. Ian Inkster gave intellectual guidance in the days when this work
was developing as a Ph.D. thesis in the Economic History Department at
the University of New South Wales. My husband was consistently supportive and encouraging. Harry Booth gave generously of his time and
mental energy in commenting on the draft manuscript. He brought to
that task his own wealth of experience in assisting the development
process in underdeveloped countries, and I have been fortunate to
benefit from it. Charles Herbert Currey, who bequeathed a fellowship to
assist the writing of Australian history from the original sources, bought
me a little space and time.

Jan Todd

Abbreviations

AAAS	Australasian Association for the Advancement of Science
AB	Adolph Basser Library, Australian Academy of Science, Canberra
ADB	*Australian Dictionary of Biography*
AGM	Annual General Meeting
ALCB	Adrien Loir Cutting Book (volume number indicated)
AMS	*Australian Mining Standard*
ANZAAS	Australian and New Zealand Association for the Advancement of Science
AONSW	Archives Office of New South Wales
APR	*Australasian Pastoralists' Review*
AR	Annual Report
ASF	Agricultural Special File
AVJ	*Australian Veterinary Journal*
BAS	Bibliothèque de l'Academie des Sciences, Paris
BN	Bibliothèque Nationale, Paris
CCRO	Cheshire County Records Office, Chester, England
CSC	Colonial Secretary's Correspondence
DT	*Daily Telegraph*
GSF	George Swan Fowler, Personal Papers, Mortlock Library, Adelaide
JLCNSW	*Journal of the Legislative Council of New South Wales*
JRSNSW	*Journal (and Proceedings) of the Royal Society of New South Wales*
JRSVic	*Journal of the Royal Society of Victoria*
JSCI	*Journal of the Society of Chemical Industry*
LA	Legislative Assembly
LC	Legislative Council
ML	Mitchell Library, Sydney

NBAC	Noel Butlin Archives Centre, Australian National University, Canberra
NLA	National Library of Australia
NSWPD	*New South Wales Parliamentary Debates*
OML	John Oxley Memorial Library, Brisbane
PC	Pasteur Correspondence
PLSNSW	*Proceedings of the Linnean Society of New South Wales*
PR	*Pastoralists' Review*
QLD	Queensland
QldPD	*Queensland Parliamentary Debates*
QldPVP	*Queensland Parliament Votes and Proceedings*
SAPD	*South Australian Parliamentary Debates*
SMH	*Sydney Morning Herald*
SSS	*Social Studies of Science*
SSSJ	*Sydney Stock and Station Journal*
TAmIMEng	*Transactions of the American Institute of Mining Engineers*
TAusIMEng	*Transactions of the Australasian Institute of Mining Engineers*
TFedIMEng	*Transactions (later Proceedings) of the Federated Institution of Mining Engineers*
TIMEng	*Transactions of the Institution of Mining Engineers*
TIMMet	*Transactions of the Institution of Mining and Metallurgy*
VicPD	*Victorian Parliamentary Debates*
VPLANSW	*Votes and Proceedings of the Legislative Assembly of NSW*
VPLAVic	*Votes and Proceedings of the Legislative Assembly of Victoria*
WAMVP	*Western Australian Minutes, Votes and Proceedings*
WAPD	*Western Australian Parliamentary Debates*

PART I
Overview

PART 1

Overview

CHAPTER ONE

Dependency at the periphery: debates and questions

To import or not to import—this has long been the question for countries not privileged to be world economic powers.[1] Many do not have any option. Australia has, in general, embraced overseas technology with a passion born out of necessity and grown into a dependence supported by a habit of mind. Was it inevitable that it should be so?

A century ago, Dr Charles A. Mulholland, Ph.D., did not believe so when he asserted that; 'It is just possible that a prophet may arise now and then in the colonies. London is not the centre of the universe.'[2] It was 1895, and Mulholland's centre was the tiny settlement of Bathurst, in the colony of New South Wales. Not only was it not London; it was 150 miles west of Sydney, the colony's capital, from which it was physically and psychologically separated by the great divide of the Blue Mountains. For two years metallurgist Mulholland had been scurrying between his laboratory and local goldmines, putting to the test his theory that the controversial new process, the cyanide extraction of gold from its ore, could be made to work better with the help of bromine. Little did he know that two British workers were on the same track. They had now tied up their results in a patent which they hoped to commercialise throughout the goldmining world. Mulholland disrupted their plans for Australian exploitation by offering his own process to the colony's miners 'untrammelled by patent rights'. As he did so, he insisted that his right to claim priority in a chemical discovery was equal to that of the British patentees. As we shall see in a later chapter, they, 'respectfully', disagreed. He was after all challenging the assumptions that still lay at the heart of British–Australian relations.

Those relations seemed to show all the signs of dependency. The British had staked their claim to Australia in 1788, with little more than a few boatloads of convicts, and the soldiers to guard them. A century on, a complex international economy had come into being with Britain as the central power, but Australia was still an infant outpost at the periphery, apparently on the edge of the world in every sense—geographically, economically, politically, scientifically. Its one saving grace was its cultural and material life-line to Britain, the 'workshop of the world'.

The industrialisation of Britain and then Western Europe was what drove the development of the international economy in the nineteenth century. These countries were hungry for all kinds of essential inputs. For the first time in history far-flung parts of the globe were connected by trade, by the flow of people, ideas, money and machines, and by a revolution in transport and communications. Australia could not help but be affected by the ferment.

From early in the century, Australia had been drawn into international trade through its ability to feed British textile mills with good quality wool. In 1850, 32 million pounds were exported from Sydney and Port Phillip alone.[3] Then came gold. In the early 1850s, the colony of Victoria produced about 3 million crude ounces each year; New South Wales added close to another 1 million, taking the total value of colonial gold output to around £12 million. That boom gradually subsided, but in 1861–65, Australian gold exports were still worth £8.7 million per year compared with wool's £5.4 million. Together, they accounted for 78 per cent of all Australia's exports. Their relative shares changed over the following decades, but by 1890 they still earned 82 per cent of the country's export income.[4]

Australia then had long since ceased to be a dumping ground for convicts. More respectable migration and settlement had been actively encouraged since the 1820s and spontaneously boosted by the attractions of gold. By that time, earlier tendencies were developing into what would become a characteristic pattern in Australia's relationship with Britain and the international economy. Australia now had a clear function in the world. Into its colonies flowed capital, labour and technology, accompanied by the transfer of institutions, knowledge and resources. Most of these came from Britain. Out of the colonies flowed exports of foodstuffs and raw materials. Most of them went to Britain. Many of the other colonies within the British Empire developed a similar pattern.

Unfortunately, from 1875 the primary products on which these countries depended were increasingly paid for at declining commodity prices, and Australia was no exception. Trade deficits reflected the import of high-priced manufactured goods and machinery; balance-of-payments problems reflected the need to borrow to pay for their purchase. By 1890 Britain was well on its way to being the 'banker of the world', and the cost of servicing the debt to Britain took more than 40 per cent of Australia's export earnings, compared with 15 per cent a decade before.[5]

Australia was then on the brink of the most severe depression in its history. It was in fact a world depression. Prices, production and employment fell everywhere, but Australia's economy plunged to unparalleled depths, and for a longer duration than most. The trigger was the flight of British capital after a leading British merchant bank, Baring Bros, faced bankruptcy as a result of defaults by Argentinian borrowers. It was not until 1907 that Australian incomes returned to their 1890 level.[6]

Another century on, Australians still shudder at memories of another depression, in the 1930s, using it as a standard of comparison for more recent economic troughs. Similarities are observed with the causes and effects of the 1890s debacle. The reliance on British inputs and markets has declined, but the features of Australia's relationship with the international economy and those who now dominate it are still familiar. It is said, for instance, that Australia lacks what has been called 'technological sovereignty',[7] and the condition seems to stretch back to the beginning of European settlement in this country. It is tempting to infer that Australia has been set in a continuous and continuing relationship of dependency from the beginning of British colonisation to today; that, though the detail of the surface pattern has changed in different periods, the structural reality beneath remains the same.

The simplicity, the symmetry, of such a thesis is seductive. Yet it conceals possible complexity. What if there were different varieties and layers of dependency? What if they did not all coincide? To explore these possibilities, we must first delve into the debates about dependence at the periphery. Then we can allow history to tell its tale.

The debates

Economic dependence

Serious consideration of Australia's economic dependence began with Brian Fitzpatrick, who had just lived through the devastation of the Great Depression of the 1930s. In 1940, after four years of intensive research, he said there was 'still something essentially "colonial" about the Australian economy, still a very effective measure of economic control by British capital interests by virtue of their investments in Australia'.[8] He went on to describe how Australia's development was conditioned by the ebb and flow of those British investments, its economy structured to suit those vested imperial interests which sought to maximise their plunder of colonial resources. This, he said, was manifest in the predominance of the pastoral sector. In this interpretation, the major collapses in the Australian economy were due to the withdrawal of British capital

and the rearrangement of its priorities: depression exposed a dependence
and vulnerability that was ever present. Imperial bonds were inherently
and structurally exploitative, ultimately designed to take out rather than
put in.

From the greater prosperity of the late 1950s, the work of N. G. Butlin
challenged this kind of interpretation in three significant ways, helped
by a new arsenal of quantitative data on the period from 1861.[9] First,
Butlin downgraded the long-term importance of the pastoral sector and
revised the meaning of the 'long boom' (1861–91), with its unprece-
dented levels of immigration, capital inflow and growth in production
and incomes. He directed attention to the urban activities of manufacturing, residential construction and provision of social utilities, showing
how they were boosted by the large inflows of British capital and labour.
These burgeoning industries, he said, gave new diversity and strength to
the structure of the Australian economy. Built around local markets, they
reduced reliance on primary exports.

Butlin's second challenge was to the idea of British domination and
exploitation; rather, he saw the relationship as a mutually beneficial partnership, in which Britain made important contributions of capital and
labour but Australians made the critical decisions according to Australian
criteria. Thirdly, with his emphasis on internal forces as the key to Australia's development, Butlin also laid the responsibility for the depth of
the 1890s depression at the feet of Australia's own decision-makers. A
satellite it might be, but one that was making its own way.[10]

From the 1960s this weighing of internal and external determinants of
Australian development increasingly took place against the backdrop of
the post-war, post-colonial world. Theories of imperialism were adjusting
to the failure of now independent colonies to achieve the economic success
they desired. In particular, dependency theory observed the unbalanced
structure of the global capitalist economy and the tendency of international
trade to drain resources away from the less developed to accumulate in
the advanced industrial countries. To theorists like André Gunder Frank,
underdevelopment was no original condition, as implied by the theory that
economic growth proceeds in stages;[11] rather, it had been created through
integration into the world economy, first through colonial administrations
and then, more subtly, through free trade and the international division of
labour.[12] Though formal colonial relations were disappearing, there was
still a centre which directed and gained from the flows, and a periphery
estranged from the sources of power.

The broad-ranging dependency which appeared at the periphery was
associated with such features as a high proportion of exports, a reliance
on primary commodities, a narrow concentration of foreign trading partners, a large external debt, extensive foreign investment, a high level of

technology imports, and unfavourable terms of trade.[13] Some writers began to grapple with the possibility that Australia fitted the underdevelopment model. Though rejected in its starkest form, the model helped generate a view that Australia's apparent modern autonomy masked an updated version of the old relationship of dependence. The idea of 'dependent development' gained currency.[14]

Scientific dependence

Science had never featured strongly in accounts of Australia's history, but when George Basalla observed recurring patterns in the transfer of Western science from Europe to the rest of the world, he drew science into the analysis of centre–periphery relations. The result was an enduring framework for viewing the growth of embryonic scientific communities in diverse geographical and social settings.[15] It was a diffusionist framework, which envisaged positive flows of resources from the centre to accumulate in the form of increasingly independent scientific activities and institutions at the periphery.

Basalla's three overlapping phases of transmission provided a timely frame of reference for two Australian historians, Ann Moyal and Michael Hoare, in their quest to lay meaning on the chronology of Australian science. Neatly fitting the characteristics of Basalla's phase 1, the period from Cook's 1770 voyage up to 1820 was dominated by the interests and patronage of Joseph Banks, operating from Kew. In the first stirring of change in the 1820s, in the boost to local science in the aftermath of the 1850s gold rushes, and then in the climactic formation of the Australasian Association for the Advancement of Science (AAAS) at the end of the 1880s, were all the signs of a progression through phase 2 'colonial science'. By the end of the century, as Australia became a federation of relatively independent states, Australian science had apparently moved into phase 3 as it came 'of age'.[16]

Basalla's model and Australian data described changes located at the periphery, but different perspectives have since changed the lens of history. In the 1980s, debate on Australian science shifted focus when Roy MacLeod also drew on the new imperialist theories and argued that Basalla had left politics out of the picture. He focused on the British Empire as a discrete analytical unit and saw what he called the 'moving metropolis'. As it extended its tentacles to dominions and colonies, the Empire used science as an instrument of policy, and shaped the growth of 'colonial science' according to its own needs.

Instead of Basalla's three phases, MacLeod saw five, and all corresponded in characteristics and timing with specific periods of imperial policy. Therefore, their characteristics and changes were of the centre, and British dominance pervaded all aspects of the relationship, including the cultural hegemony which sustained colonial loyalty. The formation of AAAS reflected a shift in the style and aims of a new stage of imperialism. Instead of an Australian national science flowering to independent maturity, MacLeod saw a scientific community estranged from its own society and acquiescent in its subjugation to imperial design.[17]

Ian Inkster injected another ingredient into the debate when he urged a different kind of analytical distinction between the countries who received European science. On the one hand were the areas of 'relative economic backwardness', where transferred science inevitably clashed with established indigenous cultures. On the other hand were the so-called regions of 'recent European settlement', those apparently 'empty' lands where the spread of science *was* the inflow of Europe's scientific community.[18] Arguing that Basalla's framework was applicable only to the latter, he acknowledged its relevance to Australia but modified it in three substantial ways.

Following Fleming,[19] Inkster claimed that dependence comes as much from the need for psychological orientation and identity as from institutional structures: Australia might be developing its own scientific societies during the late nineteenth century, but the 'mental map' of local scientists was still fixed on Britain, even into the early twentieth century. Inkster also moved behind the local indicators of development to seek the mechanisms of change. In doing so, he discarded the notion of 'science' as a monolithic enterprise and proposed instead a structure of three layers, including the socio-economic base of society; Basalla's transitions of phase could then arise from the resulting interactions.

Technological dependence

In the historical debates described above, technology appears as an ill-defined appendage. It was elevated to more prominence, however, by a third, and related, discourse about current levels of technological dependence and what it means for Australia's future.

Prompted by concern for the consequence of extensive penetration of modern multinational corporations, some observers of the Australian experience since World War II noted that rising foreign investment was accompanied by a high level of technology imports, foreign patents and external royalty payments, a low level of local research and development, and a 'brain drain' of scientists overseas.[20] The profile seemed to fit that aspect of dependency theory which saw technology transfer as the characteristic mechanism of the most recent stage of imperialism, where local industrialists link up with multinationals to gain access to their technology, but are given only part of the package. Forced to continue importing essential components, equipment, patents and technicians from the parent company, cut off from the organisational centre, they lack control of their own production process. Technological dependence appeared, then, as part of a complex set of exploitative economic relationships between centre and periphery.[21] Noting a 'technological balance of payments' which by the 1980s was annually sending around $A100 million out of the country, Wheelwright and Crough concluded that Australia was 'technologically speaking . . . an under-developed country'.[22]

When these trends were put into historical perspective, some writers drew a parallel with interpretations of continuing economic dependence. They saw a technological dependence which had become structurally entrenched in the nineteenth century.[23] Even as Australian manufacturing began to develop, the colonial relationship with Britain constrained technological developments towards the primary export sector with which British interests were aligned. It also choked off the possibility of a mature industrial base which could generate independence, particularly the capital goods industry essential to service and stimulate complex manufacturing. For Britain needed safe markets as its own manufactures came under growing pressure from other industrialising countries. Australia's continuing reliance on imported British technology then set up a reinforcing pattern, inhibiting the local development of the technical expertise and infrastructure which were a precondition for independent technological development in the twentieth century. It also tied Australia to a technology source which was increasingly falling behind the world technology frontier, and taking Australia with it.

Whether contemporary or historical, the argument from those who stress Australia's dependent position is focused on the blockages to local development; it suggests there was a missing link from foreign technology to domestic technological capacity. There is, on the other hand, another kind of argument which suggests that the consistent import of technology is a good thing. One aspect of this hypothesis points to the stimulation of complementary research and development activities in two ways: firms importing technology from overseas are usually obliged to undertake at least a minimum threshold level of research in applying it to the local situation; and the use of overseas technology by foreign subsidiaries puts pressure on domestic firms to pursue innovation to maintain their competitive position.[24] In this equation, technology transfer has not eroded or inhibited the development of local skills and infrastructure, but enhanced it.

Further support for this argument has been derived from the 'public good' characteristics of technological knowledge.[25] The fact that knowledge cannot be divided into discrete and saleable units, the difficulties of enforcing property rights, and the risk and uncertainty of research, can all undermine the incentive to invest in knowledge. When these features are applied to knowledge as an international commodity, they also point to the 'free rider' opportunities associated with technology transfer, where a small country like Australia can gain the productivity and welfare benefits of better processes and products without contributing to research and development costs.[26]

Questions

We have therefore three related but separate debates—economic, scientific and technological—concerning issues of Australian dependence. These leave us with some perplexing questions:

- Was Australia's path to federated nationhood backed with the force of some political and economic sovereignty, or did the vacillating flows of British capital, labour and technology merely emphasise Australia's role as Britain's buffer against the force of the international economy?
- Was the period from the 1880s one in which Australian science was showing all the signs of maturing to independence, or was it simply fluctuating at the mercy of shifts in British policy?
- Did local science have any direct relationship with the growing domestic technology and economy, or did it derive its impetus solely from the questions of metropolitan science?
- Did the import of overseas technology inhibit and stifle local inventive activity by substituting foreign development efforts for local? If so, did this in turn inhibit the development of a local scientific, technical and industrial infrastructure?

Unfortunately there is a conceptual gap which prevents the linkage of the three debates and the questions they raise. What follows is an outline of a framework through which these issues can be addressed in an integrated way.

The framework: seeing technology as a system

The wealth and power of Britain climaxed through the process of industrialisation which gathered pace in the nineteenth century. Those who have studied that process have made it abundantly clear that technological change was at its heart. David Landes describes it as 'an interrelated succession of technological changes': these involved the substitution of mechanical devices for human skills, of inanimate power for human strength, and great advances in the processing of raw materials, especially in the metallurgical and chemical industries.[27] Being related, these technical changes were complementary, and their interaction reinforced their effect.

Britain pioneered these technologies, but their transfer to other countries was vital in spreading industrialisation beyond Britain. To Sidney Pollard, without them, 'no Industrialization as we understand it could take place'. Yet, once they had been adopted on a wide enough basis, 'nothing appeared to be able to prevent the region concerned from "taking off"' [28] Thus was the gap between Britain and its nineteenth-century emulators gradually reduced.

Twentieth-century followers hoped that acquisition of industrial technology could do the same for them. Economic measures of technological contributions to economic growth added to the optimism.[29] But the industrialisation literature carries the overwhelming message that the technological centrepiece must be viewed within a context far broader than the machines transferred. Rather, whole societies and their institutions and resources had to be ready to receive them. Without the social, political and economic conditions to create an effective *demand* for the technology, techniques brought from Britain would have remained irrelevant curiosities. Without the human, financial and infrastructural *capacity* to put the technology into productive use, its potential would have been unrealised. Even a spinning mule transferred to Philadelphia 'confounded interested parties for four years and was eventually shipped back to Britain in 1787, leaving Philadelphians none the wiser but angrier'.[30] The coke-burning blast furnace was introduced to France in 1785, but even sixty years later was used in less than half its pig-iron production.[31] This contrasted with the rapid diffusion of the Bessemer process after mid-century, by which time the social ruptures of revolution and subsequent restructuring had begun to take effect. Technological change was embedded in a network of essential cultural, political and economic transformations. Industrial technologies were interacting not only with each other but with the whole institutional environment.

For such reasons, several theorists interested in the process of technological change have tuned into ideas about the social construction of knowledge.[32] One of their most important insights was the recognition that knowledge, including scientific and technological knowledge, does not emerge from or exist in a vacuum. It arises within a given context, through the work of people focused on particular problems, situated in particular situations or institutions, surrounded by particular kinds of peers, guided by particular norms of accepted practice. More specifically, there are many factors which go into the design and refinement of a particular technology—theoretical and empirical knowledge, resource and economic constraints, social values and tastes, legal standards and requirements, institutional and organisational structures and assumptions. Thus there is an increasing inclination to talk of technology not as an isolated physical artefact but as a social product.[33]

There have also been changes in the way technology is seen to relate to science. Once it was assumed that the facts and theories of science were handed down for application in technology. Now there is a much greater willingness to see science and technology as social and knowledge systems which interact through a two-way process, and on an equal footing.[34] The boundaries between science, technology and their environments have begun to blur and institutions have become important.

To convey all that is implied by this, it has been found useful to view technology as a system which relates its physical form to the range of factors which find expression in its specific design and operation.[35] Though system remains an elusive concept to define, there are certain essential characteristics which are at the heart of what various writers want to project. The first is the fact that a system consists of several parts. Because they are *interrelated*, the state or activity of one influences that of others; they are functionally interdependent. Secondly, 'a system is not simply a sum or heap', but has characteristics different from those of its isolated components. Thirdly, in a system the parts are coordinated, and often control is centralised. Moreover, a system has an environment which is outside its control but may influence its course.[36]

Technological systems are therefore social creations whose various constituents are brought together by individuals or organisations to solve a particular problem and to achieve a particular practical end. Thomas Edison's development of electric power has been eloquently described by Thomas Hughes in this way.[37] Most obvious in such systems are the *physical* items—from steam engines to railway tracks, from electric lamps to power plants. *Organisational* components extend from factory layout and internal management procedures to include such things as supplier firms and financial institutions. Related *scientific and technological knowledge* comes in both intellectual and institutional forms, and includes books, articles, university teaching and research programs. *Legislative* factors such as patent laws, product standards and regulation of working conditions represent social conditions and values; they have to be taken into account in designing products and can continue to influence the way technology is used.[38]

Clearly, technological systems extend well beyond the individual firms that might use them. They also evolve over time. As a technology becomes established, those in control will be attempting to widen its sphere of influence: component suppliers may adapt to its specific requirements, regulations may adjust to its standards, training curricula may integrate its new elements of expertise, organisations may modify structures to take account of its imperatives and advantage of its opportunities. Further development will thus be increasingly constrained and determined by the progressive accumulation of organisations, people and interests committed to the system, its core of technical knowledge and equipment, and all the attendant structures.

Technology transfer and technological systems

When a new kind of technology is imported from another country, we call the process technology transfer. How does it relate to these systems? Because technology arises in a particular time and place, it embodies the characteristics that suit it for use and survival in that environment. This poses problems for transfer to a location where needs, constraints and inputs may differ from those for which the system was designed. Adaptation may be necessary to take account of new political, legal, educational, scientific, social, economic and cultural institutions, as well as geography and resource conditions. Hence it is also possible to think of national systems of technology, conditioned by a unique set of historical factors, reflecting certain national characteristics, institutions, values and goals.

The import of foreign technology, and the process of technology transfer, can therefore be viewed as the coming together of two technological systems, which may or may not be compatible. Depending on the extent of their convergence, three possible types of outcome can be envisaged:

- interaction and assimilation of the foreign into the local: establishment of the link from foreign to domestic technology;
- system clash and incompatibility leading to the two systems operating in isolation and competition: the 'foreign enclave' syndrome;
- the failure of the whole process of transfer and diffusion: standard practice is unchanged, anticipated benefits never eventuate.

The relationship between sender and receiver is crucial to what follows the initial import of technology, and so are the characteristics of the receiver. Social and economic benefits will occur only when the technology penetrates deep into relevant sectors. But does the receiver have the capacity to spread information; does it have the financial structure to support investment in new areas; does it have the skills to operate the technology effectively or to carry out any necessary adaptation; do its institutions have the flexibility to adjust to new requirements? And what is the location of its control centre?

Also important to the final outcome is the nature of what is actually transferred. Perhaps only machines are imported. Perhaps a whole technological system is transferred, complete with its own organisation, finance, operators, research departments and other requisite components.

We might also ask what the cumulative effect of importing technology is. It has been suggested that a progressive process of technology transfer may effect a transition from dependence on borrowed technology to the capacity to adopt and invest in indigenous products and processes. The import of completed goods and techniques may lead to a stage where only design is transferred and technology comes primarily in the form of blueprints, formulas and books. A third stage could see the transfer of only basic knowledge, bringing the more enduring capacity to create local technology from overseas prototypes and to develop comprehensive domestic research programs.[39]

The many failures of post-war transfer and development strategies make it clear that there is nothing automatic in such a process. This book is designed to explore what happened in Australia, by means of two case studies of technology transfer in the late nineteenth century. The first is the transfer of anthrax vaccination into the pastoral industry; the second is the transfer of the cyanide process of gold extraction into the mining industry. Both industries are important parts of the export sector which has been portrayed as the pivot of British exploitation, influence and control.

Both cases also represent the introduction of new technological systems. It is now acknowledged that science and technology may advance by small cumulative steps or by giant leaps which reorder ways of seeing and doing—Kuhn's 'scientific revolutions', Schumpeter's innovative 'gales of creative destruction'.[40] In the technologies we examine here, the concepts, knowledge and requirements on which they were based did not evolve naturally out of technologies which they were to replace, but involved a shift to a different way of perceiving the problem and to a different cluster of techniques, skills and paradigms required for addressing it. As such, they represented a *technological discontinuity* and in both cases scientific knowledge was a *sine qua non* for the new technology, in contrast to the old.

Chapter 2 sets the Australian colonial scene, sketching in broad outline the receiving environment as it developed in the late nineteenth century. It was a land where the turbulent eddies of autonomy buffeted the authoritative flow of imperial control or influence. Most of Australia's institutions experienced some change as a result. This chapter focuses on those changes and institutions which were most relevant to the development of Australian science and technology, and particularly those closest to the unfolding events of our story.

The story itself is told in Parts II and III. The anthrax vaccination study is described in Part II. It is located mainly in New South Wales where the anthrax problem was largely concentrated. The technology it involved was based on the emerging science of bacteriology, and came not from Australia's traditional, imperial source, but from France. We will see whether the transfer challenges the portrayal of Australian dependence as centred on its coloniser, or affirms Britain's ultimate control.

The cyanide process of gold extraction we have already touched upon. It was developed in Britain as a method for enhanced exploitation of the mineral resources of non-metropolitan countries, among whom Australia figured as one of the most favourable for its application. This technology, then, presents us with the opportunity to explore its transfer within the context of the colonial relationship. Covering a technology which became significant in all the Australian goldmining colonies, this case takes in events in Queensland, Victoria, New South Wales, South Australia and Western Australia.

Part IV takes us on to questions of wider relevance. What do these specific cases tell us more generally about the position of science and technology in Australia during the late nineteenth century? What do they tell us about the extent of Australian sovereignty over the techniques governing domestic production? What do they tell us about the short-term and long-term consequences of importing foreign technology? What can we infer about technology transfer into European colonies like Australia?

CHAPTER TWO

Cross-currents of change

In 1883, Richard Twopeny had been an English migrant in the Australian colonies for eighteen years. As an educated and honest man, the son of an archdeacon, he felt himself well able to comment on colonial affairs. Reflecting on the baggage brought from the Old World, he noted that 'the legislative equipment of the young Australias corresponded pretty nearly to the tall hats and patent-leather boots which fond mothers provided for the aspiring colonists'. The colonies had, he said, 'imported their whole constitution and law books holus-bolus from England'. Yet he also observed that 'the adaptability and less complicated social machinery of a young colony have permitted the carrying into execution of many valuable measures long before they emerged from the region of theory in their native land'.[1]

Colonial Australia may have been on the distant periphery of world affairs, but it was, in its own way, a society on the move. Change was in the air. At mid-century, its dispersed and tiny settlements were just wresting some control over local affairs from the Colonial Office. Fifty years later, at the turn of the century, Australia was proclaiming nationhood as a united federation of states. Much had happened in between.

Gold had sparked a good deal of change, simply by tripling the population from 400,000 to 1.1 million in a decade. By 1891 the figure was close to 3.2 million. Investment had flowed as readily, with net capital imports rising from £22.4 million in 1861-65 to £98.7 million in 1886-90.[2]

The flow of both people and capital was stemmed in the depression of the 1890s but recovery eventually came nevertheless, built on what had gone before and on the increased diversification of production. By 1900 the emphasis was still on primary goods, but wheat, meat and butter were whittling away the export dominance of wool and minerals. Manufacturing was now contributing something like 13 per cent to the gross national product, and occupying close to 20 per cent of the workforce.[3]

Australia was becoming a much more complex place, certainly more than most Europeans probably realised. Most obvious in the changes was the general increase in the extent and density of settlement and production in the Australian colonies. The common pattern brought intensified concentration around the capital city, with fingers of development radiating into the hinterland. Yet there was also considerable variation in the shape and texture of development from colony to colony. Perhaps even more significant was the increasing complexity of the relationship between the imperial centre and colonial outpost. No longer was this a simple and safe dependency as the colonies sought some national identity. Part of that identity would be found in the nature and use of its technology. In this chapter we will traverse some of the changes which would determine the course that technology would take.

The political environment: shifting sovereignty

Political processes and legislative outcomes impinge on productive technology in many ways. Some are direct, as in the laws which regulate operation. Others exert their effects through the creation of a particular kind of environment in which production takes place. The introduction of colonial self-government in 1850 began a series of changes which reshaped the political environment of the country and rearranged the character of its political sovereignty.

Even from the early days, distance and local conditions could subvert the tight control of London policy and provide some local latitude in implementation. Nowhere was this more evident than in the all-important area of land policy,[4] which governed the disposal of that most precious natural resource. Early tussles undermined British plans for close settlement and by 1836 achieved the *post facto* legitimation of the wandering ways of squatting. With it came the creation of a prominent local group intent on increasing its own powers. Further turbulence in the 1840s won legal recognition of the variability of Australian land and the need for different conditions of tenure. Out of the associated ferment came the passage in 1850 of the Australian Colonies Government Act to install 'well-regulated freedom'.[5]

The Act detached the Port Phillip District from New South Wales as the colony of Victoria and gave it the kind of government by legislative council that New South Wales had had since 1842. The same was extended to Van Diemen's Land and South Australia, and was promised to what would become Queensland in 1859. For colonies considered deficient in population, political experience and sensitivity to imperial responsibilities, the 1850 Act withheld both responsible government and the control of Crown lands, but one important concession would prove crucial: that was the right, within limits, to amend constitutions. In the ensuing struggle the New South Wales Legislative Council won the essentials of responsible government before the end of 1852. In the following few years, the colonies set up most of the fundamental mechanisms of parliamentary democracy as they carved out their separate existences.[6]

But the British were not about to yield all. Sovereignty was bounded by the legal supremacy of the imperial parliament, the constitutional authority of colonial governors responsible to the Secretary of State for Colonies, and an absolute control by the imperial government of matters affecting the empire as a whole. Yet largely through a process of attrition, the 1880s saw the old system of central control withering away and by 1892 the British government retained only a few shreds of its reserved powers.[7] Federation in 1901 cemented many of the gains and marked another step toward the legislative independence which was finally formalised in the Statute of Westminster of 1931.[8]

Many writers would, of course, argue that the sway of British ideology could do what formal laws might not. Some would say that Britain's economic dominance in the free-trading international economy could achieve the same ends as political control.[9] The 1880s become highly significant then as the time when Britain attempted to reassert its imperial power, as free trade and colonial liberal policy passed their peak. At its zenith from 1885 to 1905, British imperialism aimed to rebond political relationships by means of imperial federation, to link the empire economically by imperial preferences and to organise imperial defence.[10] In isolated Australia there was some attraction to imperial defence, and to imperial preferential tariffs. But only Tasmania and New Zealand showed any real interest in imperial political federation. Indeed, despite the undoubted loyalty to Britain, there was enough dissatisfaction with the relationship by this time to make the place of Australia in the empire a matter of public debate.[11]

The spectrum of colonial opinion ranged from the imperial federationists who favoured more integrated political control of the empire, to republicans. There is some debate about where the majority of Australians stood. As Dilke said, Australian civilisation was English 'with the upper class left out'.[12] But another important factor in the equation was its changing composition, for Australia was no longer primarily an immigrant community. In 1881, of the 2,323,000 people living in the six colonies, just over 60 per cent had been born in Australia, and 34 per cent in the United Kingdom. By 1901 the Australian-born in the population of 3,771,000 reached 77 per cent.[13]

The federation of the Australian colonies at the end of the nineteenth century redefined the triangle of relationships between the colonies themselves and between the colonies and Britain. The draft constitution drawn up in Australia reflected internal forces for unity and sovereignty, and also imperial pressures and colonial fears. But it also reflected widening horizons in which America was figuring more and more. During 1889–1908, 70 per cent of the foreign affairs editorials in the leading Melbourne newspaper, *The Age*, dealt mainly with the United States.[14] America also provided a model from which to draw in preparing a federal constitution.

The Federal Constitution may not have represented any clearcut Australian culture and aspirations. Perhaps Australian federationists, more than British imperialists, did see self-interest in an enduring dependency.[15] Yet, Geoffrey Sawer has concluded that, apart from the compromise on right of appeal to the Privy Council, the terms of the federal union were 'wholly made in Australia, by Australians'.[16] The final form of the document represented a choice made by Australians, looking to their own interests, and making decisions within the range of options as they saw them, a range determined by looking further afield than British culture alone.

The establishment of the Commonwealth in 1901 was not only a claim for more local sovereignty, but an attempt to overcome the drawbacks inherent in the 'splintered sovereignty' of separate colonies.[17] However, the colonial sub-systems which nestled within the federal arch still retained large areas of self-determination. To the Commonwealth were entrusted those matters essential to any genuine union: customs and excise, defence, external affairs, immigration, posts and telegraphs, banking and currency, foreign and interstate trade and commerce, industrial arbitration beyond state limits, and taxation. But the states would still be the primary development agencies and the primary mediators between the individual and government. Their legislative domain still covered land settlement, transport, education, health, the administration of justice, general conditions of employment, poor relief, and taxation in certain fields.

Legislating for technology: partnership, protection and patents

In the life and development of these sparse colonial settlements, government had always played a major role and had generally played it in a spirit of partnership with the private sector. This showed in the public ownership of necessary but unprofitable utilities, such as transport, communications and urban amenities, many of which were taken over by

government when private enterprise failed. Private economic development also benefited from the government transfer of people and capital
from overseas.[18]

Another aspect of the partnership showed in the nature of the legislation which regulated private industry. This tended to steer, rather than
police, colonial production. Primary industries found particular guidance
in the government administrative grids that overlay their activities. For
gold production this began with the archaic English law which gave all
ownership rights to the Crown. But it was fear of the frenzy of gold fever
which precipitated the 1851 introduction of a system of licences to mine
gold. Miners' rights subsequently replaced licences, then miners gained
representation on local mining boards which eventually came under the
jurisdiction of government departments of mines in all colonies. The
departmental heads became the chief overseers of colonial gold production, operating through on-the-spot monitoring by mining wardens. The
stock industries similarly felt the paternal hand of government through
chief inspectors of stock installed from the 1860s. In New South Wales
they directed the inspectors located in the sixty-four sheep districts into
which the colony was divided. Later in the century, departments of agriculture extended the partnership to a widening range of primary activity.

Manufacturing found its own kind of relationship with government.
Partly it came through the government contracts providing base demand
which carried manufacturers between less regular, but more lucrative
work from the private sector.[19] Partly the relationship developed through
policies on protection. The argument that 'infant industries' should be
allowed to mature, safe from the competition of superior imports, went
back to the late eighteenth century, but Australian colonies had won the
right to legislate on tariffs when free trade was at its height. Taking their
cue from Britain, they essentially adopted simple revenue tariffs.
However, by the 1860s, diverging trade policies were reflecting internal
colonial conditions: industrialising Victoria stepped onto an explicitly
protectionist path which it followed for the rest of the century; pastoral-
dominated New South Wales voiced a vigorous free trade policy; other
colonies with less clearcut interests lay somewhere between.[20] Though
motivations and effectiveness are less distinct, popular arguments for
tariff walls in Victoria were clear in their intent to protect local skills as
well as local goods.[21]

Patent legislation offered different kinds of protection. In principle,
patent systems might aim to protect the property rights of the inventor,
or to use the patent as an instrument of economic and development
policy.[22] Australian colonial governments were not prepared to let technology take its course without intervention, yet they were by no means
comprehensive nor uniform in adopting the British model, which itself

shifted progressively from a bias toward property protection to a desire to stimulate both the development and the use of inventions beneficial to general welfare.

Colonial patent law began in New South Wales in 1852, right on the heels of a new British Act designed to reduce the administration costs and complexities which were such a disincentive to patent application.[23] The New South Wales Act followed the British, leaning toward a simple registration of claim, leaving assessment of technical merit in the hands of unofficial 'expert referees'. The Victorian laws required far more in the way of precise specification, and provided for publication, for opposition, for provisional protection and amendment, as well as patent extension and redress for infringement. The South Australian Act of 1859 was similar to New South Wales law, but it followed America in providing for public inspection of specifications and also demanded patent application before the invention was put into use. Later simplifications were influenced by Canadian patent laws and the local Chamber of Manufactures.

The next wave of British reform came in the 1883 Act which aimed to harness patent law to the cause of development. It tightened the examination for usefulness and validity, established the condition of 'exchange for secret' and, through compulsory licences, ensured that a worthy invention would be worked. Australian colonies followed suit, but again in piecemeal and diverging fashion. Queensland was first, adopting the form of the British Act with little alteration in 1884. Victoria did likewise and extended to twelve months the period in which amendment could be made, excluded importers from the definition 'true and first inventor', expanded the rules of revocation and established rules for infringement proceedings. Always the colony with more precise and modern patent laws, Victoria continued to update, passing further Acts in 1889 and 1890 which tightened the process.

By contrast, free-trade New South Wales was the laggard and agonised more about how far the government should go, in effort and cost, to ensure the validity of the claimed invention. Eventually, the 1887 Act cleared the way for 'struggling mechanical genius'. It reduced fees, established an examination system and administration under the authority of the Minister, and provided for publication, opposition and provisional protection for twelve months. The complete specification was to 'so particularly and minutely describe the nature of the invention and the manner of its performance as to enable any competent workman, by the light of ordinary skill and knowledge, to carry the invention into effect without experiments'.[24] Some features of the British 1883 Act were ignored, in particular the provision to allow amendment of the completed specification—a significant omission for our later story.

Australian colonial governments were motivated by a desire to encourage invention, and patent law was one instrument of policy. But how was the foreign patentee to figure in this system? Colonial law did not follow those other countries which, in the interest of establishing new industries, specifically encouraged technology *transfer* by granting 'patents of importation' or 'patents of introduction'.[25] Conversely, in contrast to American law, which in 1836 specifically discriminated against foreigners,[26] Australian legislatures were not prepared to favour *local* inventors. Australian law followed a middle course, a path adhered to as Commonwealth law overtook colonial systems in the new century.

Organising technology

Patterns of ownership of production were one influence on the organisation of technology in Australia. What stood out were extremes and changes of scale. Public ownership and operation of utilities allowed for bigness. Thus the 8500 miles of railway routes that by 1890 extended through New South Wales, Victoria, South Australia and Queensland were under the supervision of their own government instrumentalities.[27] So were services such as post and telecommunications, water supply and sewerage control.

Other areas of production developed as small, family businesses. In agriculture, land laws were designed to encourage the yeoman settler, although there was a marked increase in the size of wheat farms from around 1870.[28] Dairy farming remained a family enterprise, but increasingly fed larger co-operative processing from the 1880s. Manufacturing was also small scale, typically a family concern. Though technological changes in industries such as brewing and milling encouraged concentration in fewer, larger establishments, by 1908 two-thirds of manufacturing employees were in factories employing less than 100 people and a very large number worked in very small factories.[29]

The pastoral industry, always on the grand scale, moved gradually from the traditional ownership by individuals or family groups into the hands of corporations, ten of whom held 38 per cent of all New South Wales pastoral leases by 1890.[30] Mining began with the lone digger or small syndicate, but also attracted the large companies as capital requirements escalated and the overland telegraph brought Australia closer to the London capital market. Australians owned nearly all the mines until the late 1880s, but by 1900 British investors probably predominated through their stake in the massive capital of Western Australian mines.[31]

Financial institutions made possible and also affected the organisation of technological enterprise. In Australia they related differently to the various sectors. The financial system had developed to feed the expansion of the pastoral industry, and it maintained that bias. At least until the 1860s it also maintained a pervasive British influence: this came through the presence of British trading banks; through the introduction of British practices like branch banking, which penetrated deeply into expanding areas of settlement; and through the use of deposits to expand credit, which provided a mechanism for transferring British capital.[32] However, as local banks became more adventurous and flexible in applying these methods, especially the supply of credit through branches, they moved to the frontier of economic development.[33] Their success in attracting British deposits in the highly competitive environment existing by the 1880s ensured that British money was controlled through Australian banks almost as much as through British.[34] Another dimension was added as specialist financial institutions began to set the pace. Building societies funded house-building, while both British and Australian pastoral finance companies set themselves between trading banks and pastoralists.

In the aftermath of the 1890s depression, investment patterns changed.[35] Much more cautious British capital came through the share market, largely into mining ventures. Pastoral companies sank low, but survivors put some finance into mining and manufacturing. As British capital withdrew, the local capital market stepped in and trading banks and insurance companies released more public funds, but the generally poor provision for manufacturing continued. The banks had consistently avoided long-term loans to secondary industry and played little part in financing its purchase of capital assets, forcing it into self-financing and a reliance on retained profits for growth. Prospects changed little in the post-depression reorganisation where the new savings banks were predominantly government-owned and invested most of their resources in government securities and housing loans.

Technology as capital goods

In the industrialising nineteenth century, technology usually came in the physical form of machines and equipment used in the production process. Everyday necessities such as food, shelter, textiles and clothing were increasingly made locally, but their manufacture often required the import of semi-processed materials and finished capital goods. The Australian intake of Britain's iron, steel and tinplate exports rose from 2.8 per cent in 1872 to 9 per cent in 1878, while its share of exported British steam engines grew from 4.6 per cent in 1873 to 12.6 per cent in 1878.[36]

Australia's embryonic manufacturing sector has been described as 'narrow and unsophisticated' because it produced only light consumer goods protected by distance from Britain, and immediate inputs into railways and agricultural implements.[37] The local processing of metals and machinery seems to have been limited by its focus on repair activity, while burgeoning capital works were labour-intensive, requiring only simple equipment, offering little direct stimulus to local provision of capital goods.[38]

The absence of an advanced capital goods industry has been identified as a major gap in Australia's economy and a major feature of Australia's dependence. Yet here too there was change. Geoffrey Linge has unearthed a mass of evidence showing how manufacturing industry, including suppliers of producer goods, 'pulled itself up by the bootstraps'.[39] Engineer Peter Milner has found in the physical remains of nineteenth-century Victorian engineering industries 'a display of mature industrial skills on a scale which . . . is still breathtaking'.[40] In other qualitative data he has found that a wide range of engineering firms were, by the 1880s, producing machinery which reflected a mastery of nineteenth-century high technology. Some of them employed several hundred people.[41]

Though most machinery was at first imported, Milner believes that local foundries and engineering workshops used cheaper prices, quicker delivery, better service and an equal or superior standard of craftsmanship and design, to win their place in the market. This accords with a contemporary comment that many colonial merchants actually preferred to trade in local manufactures, even if dearer, because of the 'untrustworthiness of European manufacturers, who constantly send out articles inferior to those ordered'.[42] The transition from imported machinery to local may have been earlier for mining batteries than for boilers, and later still for steam engines. But at the end of the nineteenth century, Thompson and Co.'s order books showed that their operations in Castlemaine, Victoria, were supplying a wide range of plant and machinery, including batteries, engines and boilers, had been doing so for some time, and were not only supplying the local market but exporting as well.[43]

In Sydney,[44] Thomas Mort's Engineering Works delivered the first locomotives completely manufactured in the colony in 1870, while P. N. Russell undertook an extensive range of contracts, including substantial bridge works, and many innovative, successful and recognised designs produced by its engineer Norman Selfe. By the 1870s, the company was employing in the vicinity of 1000 workers. At the end of the century, Clyde Engineering had 450 men producing a range of machinery, including sophisticated machine tools, and was about to enter the manufacture of complex steam engines. Similar stories can be told for the works established by Robert Tulloch, and by John Heine. Heine built a thriving business around can-making machines and his own openback inclinable press, the basics of which are still in evidence today.

Added to this rising capability in the colonies was an increasing proportion of capital goods supplied from America, Belgium and Germany.[45] Importers explained to a visiting British Royal Commission in 1913 that German goods in particular often beat the British ones on price.[46] Britain had been *the* industrial metropolis. Now, with spreading industrialisation, Britain was facing a multi-polar world.

The scientific and technical enterprise

Australian scientific relations also tended to the multi-polar by the end of the century. By then science had become an observable part of colonial culture, clearly visible in major scientific institutions, more shadowy in the infrastructure which supported them.

The visible side of science[47]

Mid-century gold and self-government were what sparked the whole process, bringing people, institutions and government departments with a scientific mission of some kind. The people gathered in voluntary associations designed to foster the spirit and practice of science. Mostly they were men, but behind the scenes a few women were quietly making their own contributions to science, particularly in areas such as botany. The fortunes of the scientific societies waxed and waned, but were buoyed by the affluence and optimism of the 1880s. By 1888 they had federated within a national Australasian Association for the Advancement of Science, which took its model from the British association and hoped to provide some unified direction.

It was a time when science seemed more relevant. The universities were increasing their commitment to it, with more science Chairs, more science students and more government funding for their teaching. Governments were also expanding their own scientific services in size and scope. New branches were added as governments focused on more efficient resource exploitation. Concern with fisheries, forests and water all posed new questions for which science was increasingly asked to provide some answers, while departments of agriculture employed a range of scientific professions to take the benefits of science to the 'man on the land'. More and more engineers found their way into departments of public works, railways and posts and telegraphs. The scientific community of New South Wales grew over the three decades from 1860 at double the rate of the whole population, and by 1890 at least 20 per cent of scientists, probably many more, were employed by government departments. The depressed 1890s, however, took their toll all round, with funding cuts and falling scientific audiences.

Those audiences had come largely from the educated strata of society—doctors, engineers, lawyers, businessmen and civil service administrators were attracted by vice-regal patronage and influential leadership. Such people steered the colony's development and were interested in how science could help with that. In the 1850s and 1860s they often heard papers on water supply and drainage, transport and communications, and public health. By the 1870s, a broader menu covered the collection, description and classification of Australian natural history, phenomena and resources. It reflected the condition of a colony that was still discovering its environment and what it might provide.

By the 1880s the universities were taking a different turn, edging into research programs in biology, chemistry, physics and geology, often gripped by the puzzles of European theory. But the gaze of government scientific services remained fixed on local soil. The geological surveys charted resources from gold and coal to the waters of the Great Artesian Basin. Victoria appointed a vegetable pathologist to attack the losses caused by pests and diseases. The New South Wales Director of Forestry drove hard for the exploitation of state timbers, with the help of botanical identifications and descriptions from the Government Botanist.

The swelling numbers of government-sector scientists gave them the advantage in keeping the agenda of scientific associations trained on local priorities. This bias was not without its tensions, as Government Astronomer, Henry Russell, would learn. Having strengthened the meteorological work of the New South Wales Observatory, precisely because of its significance for the development of the country, he drew criticism for his neglect of astronomy from John Tebbutt, the self-funded astronomer whose work from his home observatory won repute around the world.[48]

Indeed, there were many competing and antagonistic forces at work on nineteenth-century Australian science. Recruitment from Britain began to be balanced by professional appointments for locally trained scientists. A certain independence of thought and direction was evident in scientific work, especially in those fields where contact with the raw data of the Australian location gave advantages of insight. Australian museums in particular seemed able to straddle international integration and local acceptance.[49] More scientists in Australia were finding their work acknowledged overseas, but also choosing to publish at home. When they did publish overseas, it was now not always in Britain. This varied with the subject, but in astronomy John Tebbutt's work was just as likely to appear in Germany as in Britain, and Robert Mathews' anthropological studies went just as often to Germany, America and France.

The overwhelming Britishness of Australian science had always contained a strong Scottish component, and had always been diluted to some extent by adventurers from other European countries, drawn by gold or exotic natural specimens.[50] But broadening horizons became more evident as the Royal Society of New South Wales sought reciprocal exchange of publications with thirty-one countries around the world. By 1895 more American institutions had responded than British, who were closely followed by those of Germany, France and Italy. German enthusiasm outranked the rest, with a 93 per cent rate of reciprocity.[51]

Yet there were conflicting messages in such data. Along with the trend to equal international exchange went the persistence of a state of mind in which the British metropolis still seemed to hold the power of dispensing approval, acceptance and acknowledgment to colonial science, and the ultimate attraction to colonial scientists. Even locally-born scientists showed a natural inclination to move on to Britain to further their training and careers, a trend reinforced by the inauguration of the 1851 Exhibition Scholarships and then the Rhodes Scholarship scheme. World War I added another force in this direction, as Australian scientists were recruited to the British scientific war effort. Though some returned to apply enhanced expertise, many did not. The truth was that it was still difficult to be at the front of scientific research in far-away Australia. For this reason, many of those British scientists borrowed to build Australian scientific institutions themselves eventually returned 'home'. The end result of these two outflows was the constant need for replenishment of scientific personnel from outside. And the departure of a key person could send a whole discipline into decline.

Cultural supporters of science

Less visible as part of the structure of science, but also less obviously dependent on the metropolis, and certainly less marginal to Australian society, the cultural infrastructure comprised organisations often regarded as ancillary to the major scientific institutions, but able to draw on a broader base of social support. The mechanics' institutes formed the backbone of this infrastructure with their object of 'the diffusion of Scientific and other useful knowledge'.[52] Beginning in the 1830s, accelerating in the 1880s, this movement took some cultural life deep into the countryside. By the late nineteenth century there were about 1000 institutes scattered throughout the Australian colonies.[53]

With their models, museums, lectures and classes, these institutions provided the general population with an easy point of access to scientific culture, and created a climate where the science of the philosophical and royal societies could grow. Along with their well-stocked libraries, came a venue and platform for science, a training and recruitment ground for science, some salaried positions, and an audience for the science of more significant institutions.[54] The mechanics' institutes, schools of arts, schools of design and schools of mines also provided 'an organizational core' on which technical education was moulded as industrial developments forced governments to extend their educational responsibilities in the latter part of the century.[55]

Other kinds of infrastructural activity also fostered a scientific culture within this otherwise hedonistic and materialistic colonial society. On a grand scale, the International Exhibitions of science and technology attracted two million people in 1888. But in most colonies, government museums, gardens and observatories had a side which functioned as a more permanent cultural vehicle. It was often through these channels that scientists reached out to society and its technical needs. Here scientists such as Archibald Liversidge could join with politicians such as Edward Coombes in establishing the New South Wales Technological Museum to help develop the colony's natural resources and build its productive base.

Invention and innovation

For its contributions to that productive base, Australian innovative activity has traditionally been described as the 'improvising-battler-against-the environment' variety. Yet many successful nineteenth-century inventors had strong technical backgrounds. The migrants included skilled artisans, but most had engineering, mechanical or scientific education grafted onto their original training. The Australians often trained themselves through mechanics' institutes and associated libraries, but later in the century they entered courses in technical institutes and schools of mines and, later still, university engineering courses.[56]

Inventive activity showed up in the growing lists of patents. Certain typical orientations were evident. Australian-born inventors tended to focus on the needs of agriculture, forestry, transport and the pastoral industry, while migrants delved into fields bearing on manufacturing, such as food canning, windpower, hydraulic brakes, vehicle headlights, and even tanks. Both trends resulted in clusters of Australian patents in areas closely related to the requirements and demands of Australian economic development—a burst of refrigeration patenting around 1867–74, a building material surge in the 1880s.

There were, of course, growing numbers of foreign patents, attracted by the commercial prospects of a developing colony. The flood of motive power patents of the 1870s was led by foreigners protecting such items as steam and gas engines, yet the figures suggest that these in turn induced Australian improvements in such items as pistons, cylinders, governors and safety valves applied to a variety of purposes. Certainly in quantitative terms, foreigners did not dominate the patenting scene. In New South Wales, foreign applications fluctuated between 30 and 35 per cent during the last two decades of the nineteenth century. In Australia as a whole, foreign patents had fallen to 23 per cent by 1918.[57]

Those foreign patents were, however, showing greater diversity in their origins. America more than Britain seemed tuned in to the specific technological needs of Australia's major fields of development, favouring the areas of food processing and refrigeration, telegraphy and electrical applications, building and mining. This complemented the shifting sources of Australia's own technical skills as migrant inventors brought skills acquired in Italy, Germany, France, America and Canada, as well as Britain.

Federating Australia was still very much a land of the British, but the world, and Australia, was becoming a much more complex place. To explore the changes in detail and to ascertain the long-run effect of importing technology, we now turn to two particular selections from the international technology shelf.

PART II
Microbes, rabbits and sheep

CHAPTER THREE

Microbes versus poisonous plants

The introduction of anthrax vaccination to Australia is a story of technology transfer into the Australian pastoral sector, and particularly into
the wool industry, one of the major staples from which Australia's export
income was earned. This dominance was enhanced in the second half of
the nineteenth century as the industry grew from 17 to 90 million sheep,
enlarged its geographical response to expanding European markets, and
adapted techniques of production to changing environments.[1]

Foremost in the changes was the surging leadership of New South
Wales, with two-thirds of eastern Australia's sheep by 1890. These were
engaged in a westward drift from the high-rainfall pastures of the east
toward the distant dry western plains. As they moved, the cheap and
undemanding grassland economy gave way to the more capital-intensive
technologies of paddocking and water conservation.[2] There was potential
for contribution from science—in pasture improvement, water management and soil fertility, to name a few areas—but as late as the 1870s,
there was little evidence of it. Disease control was one of the first areas
to soak up some science, as it struggled with the insights of European
bacteriology.

A French technological system

On 5 May 1881, crowds of strangers poured into the French town of
Pouilly-le-Fort, disturbing its usual calm. They came in a mood of excited
scepticism, eager to witness a demonstration of a new technology which,
it was said, could shift disease control from treatment to prevention.

Anthrax, or *charbon*, was a deadly disease of livestock, for which there
was no cure. Stricken animals usually died within hours of the appearance of symptoms, their carcasses bloating with black, viscid blood. The
disease had long been a threat to French agriculture, and estimates of
the annual cost of losses ranged from 20 to 30 million francs.[3] Now a

33

French scientist, Louis Pasteur, was claiming that injection of animals with his preparation would render them resistant or immune. He made bold predictions that treated animals would survive subsequent exposure to the disease, while untreated animals would die. Cynics hovered, but by 2 June the Pouilly-le-Fort trials had been declared a 'stunning success'.

The technical system at the heart of this radical new technology involved microbes, diseased blood, microscopes, culture media, incubating ovens, syringes—all the artefacts required to produce and inject Pasteur's preparation.[4] The explanation for its amazing results lay in the fact that the vaccine was a living, but weakened, culture of the microbe, or bacterium, which had been shown to cause the disease. It therefore caused only a mild form of the disease, from which the animal would recover and gain immunity against future invasions.

There were two essential steps in the whole vaccination process; one was carried out in the laboratory, the second in the field.[5] The first was a delicate operation called *attenuation* which produced anthrax cultures of reduced virulence by cultivating the bacilli (rod-like bacteria) in a suitable broth at 42–44°C. The longer the bacteria were held at this controlled temperature, the lower their final potency. Pasteur's method required the preparation of two different cultures, one of only moderate attenuation, the other considerably more affected by the heat.

The second step in the process involved the use of the two cultures on the farm, at different times. The weaker one was injected into animals to induce a mild form of the disease and a certain level of immunity. After a recovery period of twelve days, the second, stronger vaccine culture was injected. The animal could now withstand this stronger dose, and in turn developed a greater resistance to the full force of the natural and fully virulent disease.

All parts of the process were interdependent in this living system. The temperature of cultivation was critical, since attenuation relied on preventing the spore phase of the bacillus's life-cycle, through which virulence was preserved. While the bacillus could live and grow without forming spores at a temperature of 42–44°C, a rise to 45°C would kill it. The level of attenuation must also be exactly right: if the cultures were too weak, immunity would be low; if they were too strong, the animal would succumb and die. In particular, the relative attenuation of the two cultures was crucial: on the strength of the first dose depended the level of immunity developed to withstand the second.[6] Thirdly, this living system must be maintained in a pure and therefore closed state. Any contamination with other micro-organisms could introduce other disease, or retard the growth of the anthrax bacillus. Either could be fatal to the inoculated animals, so non-contaminating procedures were vital.

Masterminded by a man who would become known to history as the founder of the science of bacteriology, this technology was fundamentally shaped by its scientific elements. These were at the heart of one of the most significant debates of the time—the germ theory of disease.[7] The microscopic rods observed in anthrax blood were linked with the disease through the work of Davaine, a French scientist, in the 1860s, but in 1876 Koch, in a classic demonstration, showed that these rods actually *caused* the disease. Pasteur's decisive dilution experiment of 1877 had confirmed the anthrax bacillus as the *sole* cause.[8]

By that year, Pasteur had already brought together several components of a diverse technological system of disease control, which had ranged from diseases in beer to diseases in silkworms. For Pasteur was a master at crossing the boundaries between 'science' and 'technology', a talent which brought access to further resources.[9] Partly due to promises to relieve the annual anthrax devastation, he was unusually well supported by the state, with a laboratory at the Ecole Normale Supérieure, built at a cost of 60,000 francs, and funded by an annual research budget of 6000 francs.[10] There Pasteur had marshalled a corps of talented colleagues in Joubert, Chamberland and Roux, plus junior assistants. Beyond the laboratory, the Academy of Sciences gave him ready access to the wider scientific world through its journal and its numerous commissions and juries of adjudication.

This technological system in the making took its character from Pasteur's unique scientific perspective, which focused on the relation of a microbe to its host environment rather than directly on a disease.[11] Soon it would be defined by the principles of attenuation and vaccination. These had emerged from experiments on fowl cholera but, as Pasteur recognised their general applicability, he quickly swung his attention back to the much more economically significant anthrax.

The 1881 demonstration of the resultant anthrax vaccine was part of a well-publicised campaign to establish public acceptance of what seemed an incredible product and claim. Bruno Latour has described it as an exercise in translating between the world and language of farm and laboratory, in reformulating and enrolling the interests of agricultural groups to coincide with those of Pasteur himself.[12] Certainly, the dramatic success of the trial brought a flood of requests for supplies, transforming an annex of Pasteur's laboratory into a factory for vaccine manufacture. The returns, estimated by one source at an annual 130,000 francs,[13] financed the extension of attenuation and vaccination to other diseases.[14] At the same time the world of European science was being courted and won over to Pasteur's views of the microbial realm. And, at home and abroad came the medals and other honours which raised

Pasteur's personal and scientific profile with the public. In the words of
Pasteur's colleague, Emile Duclaux: 'It is anthrax vaccination that first
spread through the public mind faith in the science of microbes'.[15]

Duclaux also records that Pasteur had long held a vision of a research
institute which could 'transform the world'.[16] As his work moved to the
production of other vaccines, in particular the human vaccine for hydro-
phobia (rabies), the dream of a Pasteur Institute moved toward reality.
Funded by public and private donations from around the globe, the
Pasteur Institute of Paris from 1887 formed the core of this French technological system.[17]

Contact with Australian disease control

In 1881 pastoralists in Australia were blissfully unaware of the dramatic
events unfolding in France and their relevance to the Australian pastoral
industry. In this vast country, protected by natural barriers, the slowly
developing system for controlling stock disease relied on legislative measures to manage animal movement and quarantine.[18] While it incorporated
some recognition of the phenomenon of contagion,[19] it had not yet
caught up with the recent European scientific verification of the role of
living organisms as the agent of disease.

This was not to say that no-one within the system was aware of the
progress of veterinary science. Since 1863 Alexander Bruce had been at
its head in New South Wales in the position of Chief Inspector of Sheep,
made Chief Inspector of Stock in 1864.[20] The legislation which created
his position also handed him responsibility for the administration of a
new colony-wide network of disease control based on sixty-four sheep
districts, each with five directors elected annually from local sheep-
owners. Though the stock inspectors who policed the system had no
veterinary training or support,[21] Alexander Bruce had a sound Scottish
education, a family tradition of scholarship, and a personal zeal and vision
which led him to seek out the knowledge of experts wherever he
could.[22] He caught up with many advances in veterinary science during
an official trip to Europe at the end of 1872.

At that time, most of the British veterinary profession ignored this
colonial representative. But the progressive Professor John Gamgee, of
the New Veterinary College at Edinburgh, impressed him with the weight
of the germ theory of disease. Professor George Brown, of the Royal
Agricultural College and also Chief Inspector of Stock for Britain's Veterinary Department, pressed home the need for centralised disease
control by qualified veterinarians. Even more influential, the Continental

leg of Bruce's mission revealed the high standing and standards of Continental veterinary colleges and their careful scientific research on animal disease. It also advanced Bruce's understanding of several infectious diseases and measures for their control, introducing him to the concept of compulsory inoculation in the treatment of pleuro-pneumonia.[23]

Yet in the early 1880s, even had Bruce or any of the local pastoralists or inspectors read newspaper reports of the Pouilly-le-Fort trial, the significance would have escaped them. For it was not generally recognised that anthrax existed in Australia. What turned out to be anthrax was known as Cumberland disease, and was widely believed to be caused by a poisonous plant. First detected in 1847 about 30 miles southwest of Sydney, in the Cumberland County, the disease had followed the geographical expansion of the pastoral industry and by the 1880s was well established across New South Wales and Victoria, with sporadic appearances in Queensland.[24] Techniques of management varied from drenching with epsom salts to putting goats among the sheep, but the prevailing technology was to travel the sheep on the roads or to move them to the hills for the dangerous summer months. Whatever the control method, the common policy among stockowners was to suppress word of any outbreak because of the difficulty of selling stock from an infected property and the depreciating effect on property values.[25]

Though the disease had been compared to anthrax at least once in each of the three colonies, the poison plant theory prevailed and the revelations of European bacteriology were of little consequence. Even the local scientific community seemed to have made no real connection between Pasteur's work and pastoral problems.[26] But Bruce's European trip had not been wasted. By 1883 he had succeeded in having a fully qualified veterinary surgeon, Anthony Willows, appointed to his staff, the first in Australia to hold such a position. The wisdom of this move was soon to be displayed in the Riverina when the new government vet was called upon to investigate sheep mortalities in the Lachlan district. One of the farms he visited was Uarah, owned by Arthur Albert Devlin, whose family had been experiencing sudden stock deaths on their three adjoining properties. In the light of current knowledge, Willows identified the disease as anthrax, referring to the presence of the causative bacilli in the blood, and cautiously venturing the possibility of conferring immunity by inoculation with an attenuated culture of the bacillus.[27]

As the first officially accepted and publicised diagnosis in Australia, this 1883 episode started the process of translating the problem occurring under the label of Cumberland disease into one under the label of anthrax. What was still lacking, though, was a clear and direct pointer to vaccination as the answer to the problem. Willows had stressed the

dangers and limitations of the procedure and recommended more traditional measures—for individual sheep, purgatives or antiseptics, followed
by bleeding; for an active outbreak, removal of the whole mob to a new
bare, dry paddock. Perhaps he had been similarly equivocal in his
approach to stockowners. Though he elicited much information during
his investigations, whatever he said seems to have left their basic pre-
conceptions untouched and the poison plant theory intact.

The disease and its effects also prevailed, with continuing heavy losses
reaching a climax in 1884–85, at 40 per cent of *all* stock on the Devlin
and surrounding properties.[28] Despite the conspiracy of silence, some
word of the deaths leaked out late in 1885 and Alexander Bruce asked
the new government veterinarian, Edward Stanley, to investigate. At
Uarah he was shown the suspected milkweed but, despite its abundance,
found no evidence to prove that it was the cause of the reported mortalities. Rather, a post-mortem examination confirmed Willows' diagnosis,
revealing the characteristic symptoms and anthrax bacilli.[29]

But Stanley proved a more effective agent in the translation process
that was necessary for absorption of a new system of disease control.
Going beyond diagnosis, he attempted to persuade the stockowners to
accept his view of the cause of the disease. To this end he made a
practical test, drenching six healthy sheep with blood from a sheep that
had died from anthrax. Five of the six were dead within 40 hours, the
sixth soon after. His point made, Stanley left Arthur Devlin a copy of
Klein's recent text on bacteriology. Called *Micro-organisms and Disease*,
it had been published only in 1884 and contained the latest scientific
views on the nature and role of bacteria in disease.

Stanley also impressed the Chief Inspector of Stock, Alexander Bruce,
with the dramatic scale of fatalities[30] and the positive benefits of trying
Pasteur's vaccination. Though still cautious, stressing the need for 'much
careful scientific investigation', Stanley recommended a government
reward 'to induce scientists' to find a way of protecting sheep from
anthrax. While Bruce was not going to rush uninformed into rewards or
vaccination, he sought information from countries in which the vaccine
had been used. Despite precautionary testing of possible plant villains,
the thoughtful Bruce was now quite convinced of the bacterial cause of
Cumberland disease.

Step one: transferring the blueprint

The stage of searching out and screening technology for transfer now
began. Concerned with speed, Bruce by-passed the Colonial Office and

contacted the appropriate countries directly.[31] The questions to governments of Europe and America inquired about any remedial measures
against anthrax, but concentrated on Pasteur's method of vaccination:
what were the results of its use; what were the results of any attempts
to cultivate the virus locally; was it suitable for sheep, cattle and horses;
would there be any danger of introducing a European form of anthrax
by importing Pasteur's attenuated virus; were the practical results of cultivation of the virus such as to encourage its manufacture in New South
Wales; what system of culture was used; what method of inoculation was
used; and for how long were animals protected?[32]

It did not take long for Bruce and Stanley to decide that Pasteur's
anthrax vaccination promised the help their industry badly needed. By
the time the Australasian Stock Conference convened in Sydney in September 1886, answers had come from America, England and Belgium.
This correspondence confirmed their hopes but also indicated that distance and the vagaries of climate would require that the live vaccine be
manufactured in Australia: it could not simply be imported as a finished
product. For this reason it was imperative to persuade others to their
view, and the coming intercolonial gathering of Chief Inspectors of
Stock, veterinarians and stock-breeders provided the ideal opportunity
for marshalling intercolonial support.

Making the conference the venue for their first public advocacy,
Stanley and Bruce placed Pasteur's method of vaccination at the centre
of the discussion of anthrax. Stanley opened with a paper describing
the symptoms of the disease and its prevention by inoculation with a
culture of attenuated 'virus'. Although acknowledging difficulties and
hazards, he announced that Pasteur had inoculated thousands of sheep
and cattle without incident. The availability of this preventive remedy
therefore obliged conference members 'to urge the authorities to com-
mence this important manufactory'.[33] In preparation, the cultivation
would have to be studied in the Australian context, since local conditions and local breeds of animal could affect the action of the virus
unpredictably.

The response was unenthusiastic. One pastoralist admitted he had suffered 'a good deal' from the disease, but was only too well aware of
differences between the Australian and European systems of wool production, pointing to the contrast between the typically large numbers of
livestock handled in Australia and the small numbers in Europe. And
stock were so much more valuable 'at Home' than in Australia: while it
might be worth treating stud stock, the inoculation of millions of sheep
all over the colony would hardly be practicable. The fact was he
remained unconvinced of Stanley's diagnosis. He had always believed it
was due to the sheep 'eating some fungoid growth', and was adamant

about the most effective remedy—change the pastures of the sheep, moving them if possible to the hills.

The Victorian delegate, Chief Inspector of Stock Edward Curr, was impatient with discussion of scientific detail. Asserting the impracticality of a virus so apparently difficult to cultivate and so variable with climate, he proposed the usual legislative response, with compulsory destruction of diseased animals and the quarantine of those in contact with diseased animals. Tradition won the day, as the counter-arguments of Stanley and Bruce were cast aside.

No doubt it was with disappointment that Stanley and Bruce returned to their normal duties, but their spirits lifted when the French response arrived soon after. With it came a booklet on the treatment as practised by Pasteur. It is not clear whether this provided any more detail on preparation techniques than the scanty references Stanley had already surveyed, but it was the closest thing to a blueprint they had. The book was translated and used as the basis for attempting to produce a culture for protective inoculation.[34]

That Stanley had kept abreast of developments in bacteriology is evidenced by the fact that he brought Klein's 1884 text when he migrated in that year. But it is unlikely that he, having graduated in 1862, had much in the way of the skills of cultivating and manipulating bacteria. The new Government Analyst, William Mogford Hamlet, could fortunately help here. Still in Britain in 1877 when Pasteur's celebrated research was announced to the world, Hamlet himself had, as he put it, 'had the good fortune to be thrown, *nolens volens*, into the vortex of the germ theory'. Later he had been engaged in research on yeast, alcohol and beer in the brewing industry.[35] Now his duties included microbiological testing.

Even so, blueprint transfer was not an easy matter. By May 1887, a question in Parliament revealed that Stanley had submitted a proposal for the employment of a specialist bacteriologist to undertake the anthrax cultivation.[36] A logical choice was a recent arrival to the colony, Oscar Katz, Ph.D., who had been in Germany in 1885 studying Koch's methods. He was now carrying out bacteriological research at Linnean Hall in Sydney under the patronage of William Macleay, and he agreed to conduct experiments on the cultivation and inoculation of the anthrax bacillus in return for a salary of 'not less than £600 a year, and a laboratory and proper appliances provided'.[37] As will be seen, events were to overtake this offer.

While Bruce and Stanley were attempting to push things along through government channels, Stanley's efforts in the bush were taking effect. Pastoralist Arthur Devlin had been plunging into the learned depths of bacteriology. Persuaded to attribute the fatalities on his property to the anthrax bacillus, Devlin had consumed Stanley's bacteriological text and then procured other books from England, becoming 'an enthusiast' in the process.[38] Armed now with the scientific definition of the problem, and tantalised with the prospect of a scientific solution, Arthur Devlin tried in vain to induce his brothers, other stockowners and finally the Minister, to help finance a trip to Europe to obtain some of Pasteur's vaccine. Next he wrote directly to Pasteur, describing the disease and asking for a supply. To his dismay, Pasteur confirmed the advice of other overseas authorities that the vaccine was not likely to survive the long voyage to the other side of world.[39]

Rabbits to the rescue: transferring skills and materials

In the meantime, the pastoral industry of New South Wales was reeling from the effects of another more widespread problem—the explosion in the rabbit population.[40] Many methods of control had been tried, but to no avail. Introduced by a grazier in 1859, the rabbits were now eating their way across the grazing lands of Victoria and New South Wales and heading for Queensland, stripping the pastures bare as they went. In contrast to the suppressed anthrax, rabbits had a very high profile, a fact which was to help bring the skills and materials for anthrax vaccination to Australia.

Pressure for government action to stall the rabbits had been growing during the 1880s. When rewards for scalps proved ineffective, compulsory destruction by the occupants of affected land was subsidised, first by a fund raised by an annual tax upon stockowners, then by government supplements. When Francis Abigail became the responsible Minister early in 1887, he found the annual government subsidy soaring to over £200,000. Meanwhile, pastoralists were petitioning for even greater relief from their share of the burden.[41] Further away, British investors and *The Times* of London considered it 'an Imperial question'.[42] The Australian agents of the large financial institutions had been ordered to report immediately on the prudence of continuing to lend on properties in rabbit-infested areas. As the Hon. J. Creed pointed out, the disaster sure to follow the withdrawal of these millions could only be averted by finding a really effective method of rabbit destruction.[43]

This was to be a search for technology on the grand scale. The cost and political profile of the rabbit problem was such that in August 1887 the New South Wales government offered a prize of £25,000 for the successful demonstration of a method to exterminate the rabbits. To

maximise the search effort, details of the prize were advertised in newspapers around the world.

Such bold action brought a bold and ambitious response. As Louis
Pasteur scanned the pages of *Le Temps*, he pondered the notice from a
distant land and believed he could help. On 27 November 1887 he put
pen to paper in a letter sent through the same newspaper. It suggested
that an infectious disease, specific to rabbits, could be used to exterminate the pest. Then, with the help of his nephew and assistant, Adrien
Loir, Pasteur put theory to the test. The first experiments were done in
the laboratory, then repeated in a farmyard. By 28 December 1887,
Pasteur was convinced that his microbes of chicken cholera could solve
Australia's problem.

Two days later Pasteur revealed his desire to send two assistants to
conduct large-scale trials in Australia.[44] One attraction was the prize of
£25,000, which would swell the coffers of his beloved Pasteur Institute.
As well, his mind was dwelling on the largesse which would come from
grateful pastoralists.[45] Although he had ignored earlier pleas for assistance
from New Zealand,[46] Pasteur now accepted the advice of the New South
Wales Agent-General in London that he should send his delegates at his
own cost to demonstrate the method on the spot.[47]

No time was lost. Pasteur's representatives were to arrive early in 1888.
Once there, they could also respond to Devlin's queries about Cumberland disease,[48] whose ruinous effects had been confirmed by the Agent-
General. With some relief, Alexander Bruce and Edward Stanley set aside
their so far unsuccessful efforts to have the vaccine produced locally.
Bruce advised the Minister to take advantage of the Pasteur mission by
arranging for a qualified local person to be taught how to cultivate the
anthrax bacillus.[49]

Pasteur's delegates arrived in Sydney during April 1888. His nephew,
Adrien Loir, still studying to become a medical doctor, had been an
assistant in Pasteur's laboratory for eight years. To help smooth the
way in Australia, Loir carried a letter from Agent-General Daniel Cooper
to Premier Henry Parkes, asking that he be given a French-speaking
assistant and introduced to 'those who will show him kindness and
attention'.[50] His companions were Dr N. Germont, who had worked
with Pasteur in his laboratory for one year, and a young Englishman,
Dr Frank Hinds, who had been with Pasteur for a mere ten days before
leaving Paris but had agreed to accompany the Frenchmen as their
interpreter.

Alexander Bruce was quick to move. On 24 April 1888 he presented
a letter on behalf of the Minister, seeking Loir's terms for instructing a
government nominee on Pasteur's methods of cultivating and inoculating
the anthrax bacillus.[51] But Loir was first obliged to carry out Pasteur's

instructions to verify the identity of the disease and assess its significance. Accordingly, on the evening of 30 April, Germont and Hinds left for Devlin's property, accompanied by Edward Stanley, to conduct the necessary tests. In preparation, Devlin had that day infected three sheep with Cumberland disease. Next morning one of these died, and Germont and his colleagues carried out a post-mortem examination in a small building set up as a laboratory. The preliminary diagnosis of anthrax was strengthened when the other two sheep died and were examined. Then the blood of one of the dead sheep was inoculated into two lambs, one of which died within 40 hours with the typical lesions of anthrax. Back in Sydney, further diagnostic tests carried out in the laboratory of the Government Analyst confirmed that the dreaded Cumberland disease was indeed the European killer anthrax.[52]

Devlin, Bruce and Stanley now initiated a campaign to generate support for a trial of Pasteur's method and stimulate demand for the vaccine. In this they would use all the networks provided by the system of disease control. Taking advantage of Germont's presence, they began with the local Narrandera Sheep Board, where Devlin, as chairman, introduced the Pasteur delegation and Stanley to his colleagues. They were obviously impressed. After a full discussion, they resolved that the government be 'strongly urged to provide a laboratory for the purpose of cultivating vaccine for preventive inoculation for the disease known as Cumberland, which disease the Board believes to be identical with that known in France as Fierre Charbonneuse'.[53]

In a more public initiative, Devlin wrote to the editor of the *Sydney Morning Herald* about the French visit, pointing out that Pasteur's representatives also brought the knowledge of how to cultivate microbes of anthrax, a far greater enemy of the pastoralists than the rabbits. He asked that the government be pressed to establish a laboratory for the cultivation of the vaccine and to conduct a public trial. He sent a similar circular to the Stock Department and to all the stock boards, and also approached Henry Parkes, to whom he was personally known. Bruce also did his part: with the completion of the Frenchmen's report to the Minister, he appended a minute recommending a cablegram be sent to Pasteur seeking the terms on which he would allow his representatives to pass on their skills. Eager to publicise the matter whenever possible, Bruce immediately made the report and minute available to the press[54]

On 23 May came Pasteur's reply, via the Agent-General. He would despatch vaccine for a demonstration at once, ensuring its stability by sending the attenuated culture in the highly resistant spore form. On its arrival, his representatives could prepare the appropriate cultivations for inoculation and also instruct someone in the preparation and use of the 'seeds or spores of the vaccin de Paris'.[55] His terms had two main components: the government should repay the expenses of his representatives and finance the establishment, maintenance and staffing of a laboratory; the vaccine itself would cost 2d for small animals like sheep and 4d for large animals like oxen. Pasteur claimed that, despite some coaxing to demand a large lump sum, he asked only what other countries had agreed to pay.

The Chief Inspector of Stock recommended immediate acceptance, but the Minister clearly wanted to test the level of demand by consulting the stock boards on several points—whether to accept the offer, how costs should be allocated, and whether a staff and appliances for supplying vaccine should be provided for.[56] The opinions of the stock boards were divided but in general favoured acceptance of Pasteur's offer. Of the sixty-four boards, thirty-eight were in favour and only thirteen against; four declined to express an opinion, four were doubtful, and the remaining few did not reply. In Narrandera opinion was favourable when a public meeting of pastoralists on 14 June expressed formal appreciation of the services of Pasteur's delegates as described by Devlin. Devlin also did his best to convince Wagga pastoralists of the virtues of vaccination, with a detailed presentation to the Wagga Stock and Pasture Board.[57]

There was greater divergence in stock board opinion on the *terms* of Pasteur's offer, and on *who* should pay the costs of transfer. Bruce's circular had canvassed several options. Now the Corowa Pastures and Stock Protection Board proposed that Pasteur should be paid a lump sum of £50,000. The Tamworth Board, still not convinced of the diagnosis, felt that Pasteur should prove the value of his methods before any public money was spent, and even then those stockowners wanting the vaccine should incur the costs through private arrangement with Pasteur. Between these two extremes sat the Kiama Board, prepared to sanction government purchase of Pasteur's method but only after thorough government testing of the efficacy of the vaccine, and only after Pasteur had taught his process to a Ministerial appointee. Then it would be appropriate for the government to establish a laboratory to supply the vaccine.[58]

To dispel the doubts apparent in some replies Bruce quickly produced another circular providing evidence that the disease was both infectious and contagious and therefore a threat to *all* owners of stock. By 28 July 1888 the Minister must have been satisfied of sufficient support to justify a public trial. On that date Bruce requested from Germont the details of the proposed test program. For publicity purposes, the correspondence was again provided to the press, prompting reports that the experiment was 'warmly approved of by the Minister for Mines (Mr. Abigail), who describes it as being of the greatest importance'.[59]

In the meantime the vaccine material had arrived, delivered to Loir on 18 July. The Frenchmen prepared a vaccine culture at the laboratory which had been set up on Rodd Island, in Sydney Harbour, for experiments on rabbit extermination. At a private trial on Devlin's property in the presence of Stanley, the vaccine proved entirely successful and plans went ahead for the public demonstration.[60] But the stormclouds of conflict were gathering on the horizon.

CHAPTER FOUR

Contagion, conflict and compromise

In the Australian colonies, anthrax vaccination interlocked with rabbit extermination, and initially this was an asset. The government's bold action against rabbits brought Pasteur's knowledge and product quickly to Australia with his own trained representatives—a great acceleration to the process of transfer. However, the initial advantages of this link soon dissipated in the face of numerous points of conflict which surfaced as the rabbit affair progressed. As tension mounted, Pasteur refused to permit any action on anthrax vaccination until the rabbit matter was satisfactorily resolved.

Systems in conflict

The opposition which incurred Pasteur's wrath came from several sources. The first arose from the complexity and conflicting interests of what might be called rabbit politics. The fact was that there were those who had much to gain from the rabbit invasion: the 'rabbiters' commonly made £10 or £12 a week, and were said to have done all they could to destroy the natural enemies of rabbits; in the leased and less productive zone of the Crown lands the rabbit subsidy could also be more profitable than the sale of wool, especially as wool prices fell. But landowners in the richer regions had nothing to gain. They supplemented wool income by supplying the British refrigerated meat trade, and wanted an end to pasture-consuming rabbits.[1]

In the political turmoil which resulted, Minister for Mines, Francis Abigail, lost the administration of rabbit control, which at the end of 1887 passed to the Minister controlling the rental of Crown lands. This meant Abigail also lost the program for finding a method for rabbit destruction. With the administration of stock disease, hence anthrax, still in his Mines Department, the bureaucratic responsibilities and attitudes to Pasteur were divided. As the Lands Department set about making preparations for a Land Bill and a Rabbit Bill which would grant concessions to Crown land lessees, it was subject to all the forces which could use the rabbit problem as the basis of their case and who consequently were ready to deprecate the value of Pasteur's method. The position was developing where the positive and enthusiastic response and advice coming from the Stock Branch was in conflict with an increasingly negative response and advice within the Lands Department.[2]

There were also those who were genuinely fearful of the repercussions of using disease as proposed by Pasteur. The poultry industry protested that the introduction of chicken cholera endangered the whole poultry population. Others were afraid for their own health, particularly after news that a disease in cows could be transmitted to humans. One correspondent to the newspapers was adamant that the discovery should highlight 'the absurd insanity of introducing these germs'. An editorial in the *Sydney Morning Herald* was convinced the microbes would kill the rabbits, but warned against playing with 'these live poisons', and asked 'what else will they do?'[3]

Fuelling the fears was a local scientific community affronted by Pasteur's apparent disregard for an unknown environment and quick to perceive an opportunity to enhance its own role. Against Pasteur's allegedly cavalier proposals, four major points were raised: the improbability that the method would be effective, particularly under Australian pastoral conditions; the possibility that it would be dangerous to humans or to animals, including native fauna of which Pasteur had no knowledge; the potential for setting off a disastrous ecological chain of events; and Pasteur's reputed poor record in applying laboratory results to the field. There was no evidence of readiness to accept Pasteur's assurances that all would be well.

Prominent among these critics was the Board of Health, attached to the Colonial Secretary's Department to advise on and administer issues of public health. Consisting principally of leaders of the medical and scientific community, it was the logical source of advice on bacteriological matters. The first blow was dealt by its Chief Medical Inspector, Dr Ashburton Thompson, who warned early in February that 'Pasteur has not been so happy in applying to man or to animals his methods of bacteriology.'[4] The second blow was delivered when the board itself resolved formally to reject Pasteur's proposals on the grounds of unpredictable danger and the lack of evidence of effectiveness. It cautioned that if the results were harmful, either to humans or stock, 'it would almost certainly be beyond the power of any authority to remedy the mischief which had been done'.[5]

A similar tone pervaded advice from Dr William Camac Wilkinson, lecturer in pathology at the University of Sydney and also the MP for the Glebe electorate. Wilkinson had worked for a period with Robert Koch in Berlin,[6] and in a letter to the Secretary for Lands in February 1888, made clear that he was not at all disposed to accept Pasteur's word that 'we shall have all good and no evil results'.[7] Concerned about the unpredictability of the microbes in a different environment, he advised that until investigations by local scientists had demonstrated that fowl cholera would destroy rabbits without doing any other harm, its introduction into Australia should be forbidden. The message was repeated from Victoria. In the newspapers Dr Weber described how chicken cholera microbes killed mice, fowls, pigeons, in fact all poultry, and would probably also kill cattle. Then, at a well-publicised meeting of the Royal Society of Victoria, Dr Henry Wigg's paper foretold how chicken cholera spores 'might be scattered broadcast in dust storms from one colony to another without losing any of their virulence'. Indeed, he suggested their virulence would be even greater on Australia's virgin soil, with new species and a totally new environment. He foresaw the death of domestic fowls and native insectivorous birds, followed by a plague of grasshoppers and locusts that might prove worse than the rabbit pest. Unimpressed with Pasteur's results with rabies vaccine, Wigg recommended that the introduction of the microbes be made illegal pending full investigation, a stand supported by formal resolution from the Royal Society of Victoria to all Australian governments.[8]

With such scientific advice, it is not surprising that the New South Wales Minister for Lands, Thomas Garrett, was panicked into the typical legislative response. Learning that Pasteur's nephew was on his way with some of the offending organisms, he rushed through Parliament a bill to arm the government with powers of control.[9] By 20 March, a few weeks before the Frenchmen arrived, this new legal weapon had been formalised in the provisions of the *Animals Infectious Diseases Act 1888*, drafted with the assistance of the Board of Health.[10] Now, without a licence from the Minister for Lands it was unlawful to introduce, keep, propagate, distribute or inoculate 'any microorganism, mite, or parasite, or other minute form of life, capable of producing any disease, or pestilence in man or in any wild or domestic animal, or any substance or medium impregnated therewith'.[11]

The effect of the legislation was to oblige Pasteur's representatives to make special application to the Minister before they could even land their microbes in the colony of New South Wales. They were therefore delayed for some time in Melbourne while the application was processed, and were feeling frustrated and intimidated before they even reached Sydney. Adding to the sense of ill-will was a letter from a Frenchman resident in Australia, who wrote to Pasteur warning that the Australians would do anything to secure the procedure for themselves, were extremely greedy for profit, and jealous of everything foreign, especially

French. This correspondent also informed Pasteur that since the first revelation of his letter to *Le Temps*, there had not been a single issue of an Australian newspaper which did not include a hostile article on Pasteur's proposed method.[12]

Pasteur warned Loir of what to expect, but predicted that these problems would gradually settle down.[13] Nevertheless, he set the tone that was to cause so much trouble for his representatives during their stay in Australia when he laid down Loir's two principal objectives: to demonstrate the efficacy of the disease for killing the rabbits and its harmlessness for sheep, cattle and horses; and to be uncompromising in asserting that these demonstrations met the conditions for the prize of £25,000. With regard to the large-scale application of the method, Loir was to say only that Pasteur had a practical plan all prepared.

In the meantime the government had made another typical response by accepting the recommendation of the Minister for Lands to establish a commission to inquire into the use of contagious disease to exterminate rabbits. The commission and its membership was to become the greatest of all provocations to Pasteur. Though the appointment of unprejudiced overseas experts from both the Pasteur and Koch schools of germ theory had been advocated in some quarters, the government clearly wanted the investigation to be a local matter, and on 10 February 1888 announced that it would appoint an intercolonial commission composed of the most experienced experts in each of the Australasian colonies. A commission of European experts, the Minister said, would take too long to arrange.[14]

The New South Wales members of the Rabbit Commission included many eminent scientific men. To Loir it appeared that the various colonies had deliberately chosen as representatives the people most openly hostile to Pasteur's project.[15] He recorded that one member was president of the association for breeding poultry, another imported the rabbit-proof fencing being erected all over the countryside, while the chairman of the Experiment Committee, Dr Wilkinson, was a follower of Koch, at that time Pasteur's rival. To add insult to injury, the person appointed as bacteriological expert to the commission was a German. This was Oscar Katz, who had formerly offered his services for the preparation of anthrax vaccine. Added to this, Loir saw strong evidence of Australian nationalism and hostility to a delegation consisting of two men who could not speak English and a third who was an Englishman, the ultimate insult.

The commission was certainly determined to set its own agenda and not to be dictated to by Pasteur. Headed by the president of the Board of Health, it met for the first time on 16 April 1888 and on 24 April received a written statement of Pasteur's scheme, including a list of three demonstration experiments. The first two would show in the laboratory that rabbits would be killed by the addition of chicken cholera microbes to their feed and that domestic animals were not affected. The third experiment would be conducted inland on an area of about 500 acres, enclosed with rabbit-proof fencing but otherwise under natural conditions. Infected food would be placed in different parts of the area, fresh broth being prepared each morning and evening to avoid loss of potency on exposure to air and sun, thus replicating Pasteur's plan for a depot in each district to supply fresh broth for distribution by pastoralists.

The following day the commission determined its own schedule of experiments, designed with a different slant. These would test whether chicken cholera would *spread* from infected rabbits to healthy rabbits in hutches, cages, and artificial burrows; whether domestic animals were at risk; which birds were susceptible; and whether the virulence of chicken cholera was altered when it was repeatedly communicated from rabbit to rabbit.[16]

The impasse with Pasteur arose to a great extent through Pasteur's ignorance of Australian conditions. Some pastoralists visiting Paris in February 1888 had astonished him with their revelation of the Australian scale: when one man said he had 3 million hectares and 1 million sheep, Pasteur exclaimed: '*C'est effrayant*' (It is frightening).[17] But this did not deflect him from the idea that a factory would be created in Sydney to prepare culture for distribution by stockowners. With this in his mind, Pasteur refused to concede any justification for the commission's insistence that experiments must be conducted to test the *communicability* of chicken cholera from rabbit to rabbit, before any inland experiment could be permitted. 'Why put so much emphasis on contagion?', he queried.[18]

In Pasteur's mind it was enough to prove the success of the method on 40-50 hectares by proceeding 'burrow after burrow, field after field', watering the grass with the culture in the evening when the rabbits would emerge to feed. Clearly he still had no real conception of the impracticability of a procedure which relied on hand feeding in the expanse of the Australian pastoral industry. As an anonymous member of the commission later put it: 'M. Pasteur, arguing from the conditions of a French backyard to those of a continent, had unwittingly deceived himself, and had claimed the great prize in a mist of perfect ignorance as to what the rabbit pest in Australia, Tasmania and New Zealand really was.'[19]

For this reason, Pasteur was reluctant to agree to the commission's demand to prove contagion, *independently* of the administration of infected food, until his inland experiment had been completed. This hiatus first surfaced on 14 June when the delegates sent a long telegram

THE PASTEUR RABBIT PROCESS.

Figure 4.1 The *Tribune*'s impressions of Pasteur's method of rabbit extermination (*Tribune*, 12 April 1889)

to Pasteur, relaying commission requirements: they were to perform an experiment testing the contagiousness of the disease, and they were also to provide infected rabbits for special experiments by the commission.[20] Pasteur's response was considered unsatisfactory, while Pasteur began complaining bitterly about the commission to British friends and French and Australian officials.[21]

As relations deteriorated, Pasteur adopted the stance of using anthrax vaccination as some kind of bargaining lever with the New South Wales government. By 23 June Pasteur felt relationships were on the brink of collapse, and on 28 June he arranged for publication of a letter which contained the first hint of his hardening attitude. While dismissing the rabbit contagion question as of minor importance, he stressed that he alone held 'the secret' of anthrax vaccine manufacture and that it would affect the future of Australian agriculture.[22] A few days later Pasteur responded to news that anthrax demonstrations could begin on 10 July with a telegram informing Germont that the steamer *Oceanen* was carrying the vaccine and a letter to Parkes; but he must not make any anthrax experiments until the rabbit question had been settled.[23]

For a time, distance thwarted Pasteur's intentions to delay the transfer of anthrax vaccination. An explanatory letter of 11 July took five or six weeks to reach Loir. For reasons unknown, the preceding telegram, dated Sydney 18 July, was not delivered to Germont until 23 August, by which time he and Loir had begun their private trials and were committed to the public demonstration of the vaccine. Given the contents of the telegram, there were suspicions about the reason for its delayed delivery. These were intensified by the fact that the letter accompanying the vaccine, sent on 30 May under cover of Sir Henry Parkes, was delivered to Germont already opened, and several days after its arrival in Sydney.[24]

In the meantime though, in ignorance of Pasteur's current attitude, the Stock and Brands Branch of the Department of Mines was keen to maintain the momentum and, having received Germont's plan for the public trial, was making further arrangements. On 9 August the government appointed the Anthrax Board to observe and report on the demonstration of Pasteur's vaccine. The composition of this board provided a stark contrast to that of the Rabbit Commission. There was no opposition here. The Anthrax Board members were the major proponents of the vaccine. To the inner cabal of Alexander Bruce, Arthur Devlin, Edward Stanley, and William M. Hamlet was added John Lamb, chairman of the Board of Sheep Directors, Sydney. He had established friendly relations with Loir at the Union Club, and had convinced Loir of Abigail's sympathy to Pasteur's mission amid the general hostility encountered by the Frenchmen on their arrival in the colony. He had indeed been instrumental in bringing Loir and Abigail together during those stressful days.[25]

Objections flowing over from the rabbit controversy were quickly quashed as Bruce responded with vigour to claims from the scientific community that there was no secret and that at least six people in the colony were fully capable of translating the bacteriological knowledge of the day into technological outcomes.[26] At the same time preparations were made for the conduct of the public trial on land about a mile from Junee Junction, near Wagga Wagga. By the time Germont and Loir received the ill-fated telegram, on 23 August, arrangements for fencing and an improvised laboratory were well advanced. This presented a dilemma, especially as Loir had, in a letter dated 1 August, conveyed to Pasteur their own concern about his specified contagion experiments.[27] To add to the confusion, Pasteur had by now authorised his representatives to hand over rabbits infected with chicken cholera for the conduct of the Rabbit Commission's own schedule of experiments.

The 'scientific proof'

A point of no return had been reached. By the end of August all was ready for the demonstration which was intended once and for all to translate the ravages of 'Cumberland disease' into scientific terms. Here the pastoralists of Australia would see the merit and indispensability of Pasteur's scientific solution. On 3 September the Anthrax Board proceeded to Junee to witness the first vaccination. The next day the first inoculations with Vaccine I began on sheep and cattle purchased from a district free of anthrax. Twenty sheep received 0.12 mL in the thigh and four cattle 0.25 mL behind the shoulder. Their temperatures were recorded for three days, that of the sheep rising on 5 September but falling the next day. The vaccinations were performed mainly by Germont, supervised by Loir, and observed by all members of the Anthrax Board. Nineteen sheep and two cattle were left unvaccinated to act as controls.[28]

A fortnight later, on 18 September, the same quantity of Vaccine II was given to the animals, which had in the meantime been held in the same paddock as the unvaccinated controls. On the same day, Germont inoculated two sheep with the virulent anthrax culture isolated in May. This culture had since been kept under lock and key and was used now to eliminate any suspicion of a French culture having been used in the experiment. These sheep died within 32 hours and were carefully post-mortemed by Stanley and Devlin in the presence of Germont and Hamlet. Under the microscope their blood revealed the characteristic bacilli of anthrax. Hamlet returned to Sydney with samples of the blood and there, under his supervision, Loir prepared the culture for the final test.

This challenge culture was taken back to Junee on 29 September. The next day it was used to inoculate three sheep, which died 34, 43 and 59 hours later. The blood from one of these sheep was used for the actual challenge on the afternoon of 2 October, when vaccinated and unvaccinated sheep were inoculated alternately from the same syringe, observed by all the members of the board. Also present, by invitation, were Dr Ashburton Thompson, Chief Medical Inspector for the Board of Health, the Chief Stock Inspectors of Victoria, Queensland and Tasmania, and the Government Veterinarians of Victoria and Tasmania. There were also several representatives of stock boards, a good many inspectors of stock and interested visitors from all parts of the colony; in all, an audience of about 200 people. Some remained for long periods; others came and went as the tests proceeded.

On the evening of 2 October, the official interpretation of the trial began with a meeting convened in Crawley's Hall, Junee. Here the audience heard lengthy papers by Edward Stanley and William Hamlet.[29] Stanley described the history of the disease, its nature, causes, symptoms, pathology and treatment. Its scientific definition was summarised in his statement 'The anthrax bacillus is ... the criterion of (anthrax) Cumberland disease.' Hamlet followed up with more scientific detail—the history of the research on the anthrax bacillus, the growth characteristics and requirements of the organism, the method of attenuation used to produce the vaccine, and, the ultimate proof of the link between Cumberland disease and the anthrax bacillus, the applicability of Koch's postulates, the four criteria that must be met before a disease can be attributed to a specific organism.

On the next evening, a second meeting was convened. At 8.15 p.m. the first unvaccinated sheep died. The rest were to follow over the next 33 hours. As a veterinarian, Tasmania's Archibald Park stated that the experiments had proved conclusively that anthrax could be dealt with by the cultivated vaccine and he believed Pasteur's vaccination would prove a preventive; the poisonous weed had nothing to do with the disease. Professor Kendall of Victoria was also warmly supportive and strongly advised all owners of infected runs to vaccinate at once. When one observer called for a larger demonstration of efficacy, Edward Stanley reassured him that the present demonstration was quite adequate. Nevertheless, it was announced that the Anthrax Board had a large-scale test under consideration and had already been offered 5000 sheep for the purpose.

By 5.30 a.m. on Friday, 5 October, all nineteen unvaccinated sheep had died, while all the protected sheep and cattle were unaffected. Stanley's post-mortems on the dead sheep found the lesions of anthrax in every case. During the examination, physiological changes and other points of interest were explained to the visitors. A healthy sheep was killed to compare its organs with those of a diseased sheep, and the blood of each examined and exhibited. To remove any possibility of doubt as to the cause of death, the board asked Hamlet to examine the blood of the sheep as the post-mortem examinations were made, and in every case the blood viewed under the microscope contained the anthrax bacillus. Finally, a sheep which had been vaccinated and challenged was killed, but showed no trace of the disease, nor any effect on the wool.

Having acquired the appropriate concepts, labels and terminology from Stanley and Hamlet, spectators were now able to witness the corresponding phenomena. The demonstration was pronounced a complete success. Anthrax Board members were 'unanimously of the opinion' that Pasteur's agents had conclusively demonstrated the efficacy of the vaccine, and recommended its adoption. The audience of visitors was similarly impressed. Victoria's Edward Curr, formerly impatient with scientific detail at the 1886 stock conference, was thoroughly satisfied. Thomas Tabart, Tasmania's Government Veterinarian and representative on the Rabbit Commission, also admitted that he had gone to Junee 'very sceptical as to the results', especially as the rabbit experiments at Rodd Island were not successful. But he was convinced that the experiments at Junee 'were conclusive'.[30]

Meanwhile, Pasteur was planning a strategy for transfer of the anthrax vaccine technology to Australia: timing was critical. In a letter arriving after the Junee experiments had begun, he insisted that nothing be done until the full demonstration had been pronounced a success.[31] This would be the time to deliver conditions to the government in Sydney. As the unvaccinated sheep started dying, Alexander Bruce cabled to his Minister, Francis Abigail, to come to Junee to witness this triumph of science. Descending from the train on 4 October, the Minister could not conceal his delight. As instructed, Loir seized the opportunity. A short conference produced an agreement that Loir would be installed in a laboratory, that the government would pay for assistants, and Loir could cultivate the vaccine to sell for 3d per head (not 2d as previously stated), on the condition that the pastoralists would offer a minimum of 200,000 sheep for vaccination.[32]

At a public meeting hastily convened by Stanley, Abigail asked the stockowners to support this pact by offering their sheep. Loir later described the following ten minutes as one of the most moving scenes of his career; from all sides of the Junee Junction station, calls came forth of 3000, 6000, even 20,000 sheep. Counting revealed there were 260,000 sheep to vaccinate.[33] To complete the festivities, the 'Committee of the Anthrax Demonstration' made Loir a special guest at a function at the

Refreshment Room at Junee. Back in Sydney the next day, the Minister passed on the feeling of triumph in a letter to Premier Parkes. The experiments, he said, had been conducted 'with the most wonderful results. Men of science who came to curse saw and went away blessing.'[34]

A change of heart

The prospects for anthrax vaccination were looking good. The stockowners accepted the asking price and the government was happy to establish and fund manufacturing facilities. Germont cabled Pasteur on 11 October for authority to carry out the vaccinations. But Pasteur's reply carried new terms. For a guaranteed supply of the spores of anthrax vaccine from the new Pasteur Institute, he was now asking a lump sum of £100,000. This would include the services of a young 'savant' trained by the institute, but all material expenses would be charged to the government. Germont was to await settlement of these conditions before proceeding with the first practical application.[35]

Despite the success of the Junee demonstration, the prospect of paying out £100,000 put a completely different complexion on the government's assessment of the whole scheme. The prompt rejection of these terms caused Loir and Germont to recommend that Pasteur revert to his original conditions laid down in May. Pasteur compromised. First they must finish the rabbit experiments. Then, and only then, they could vaccinate the 200,000 offered sheep, on the conditions agreed on in October.[36]

The temporary reprieve provided by communication problems faded away as once again the transfer of anthrax vaccination became conditional on the prior settlement of the rabbit dispute. But resolution was becoming less likely. On 1 August Pasteur had written to authorise Alexander Bruce to publish a letter confirming that Pasteur alone possessed the secret of anthrax vaccine manufacture; in it he appealed to the Australian public to seize the opportunity against the bluff of self-interested parties. More provocatively, Pasteur was adamant that he alone held the means for large-scale use of chicken cholera to exterminate rabbits, and nothing would be revealed to the Rabbit Commission. Only to the government would he make this known, and only if the prize of £25,000 was first awarded to him. The commission interpreted this as a declaration that Pasteur 'expects the reward to be adjudged him for his scheme on such proofs of its efficacy as he may determine, and before the mode of carrying it out on a large scale shall have been revealed'.[37] This, they said, was contrary to the terms of the proclamation concerning the reward.

If relations with the Rabbit Commission were reducing the prospects for anthrax vaccination, so was an emerging dissenting interpretation of the meaning of the Junee experiment. On 12 October the Anthrax Board presented its favourable report to the Minister. Soon after, opposition surfaced in the Parliament, beginning with veterinarian John Stewart's query as to the status of Pasteur's vaccine both with veterinary schools overseas and with the New South Wales government. It was a line of questioning which continued until 1890, advancing the argument that the method was neither efficacious nor safe and that there was some conspiracy abroad to infer legitimacy by government endorsement.[38]

Influential pathologist and MP Dr Wilkinson rejected the Anthrax Board's interpretation of Junee when he addressed the Australasian Medical Congress in January 1889.[39] Asserting that 'the demonstration found us where it left us', he argued that since the protective capacity of an attenuated culture had already been established, the outcome of the demonstration, in the hands of Pasteur's own pupils, could be taken for granted. But this reflected only laboratory conditions and did nothing to meet a host of objections against immediate application in the field. These ranged from Koch's counter-findings to unexplained fatalities, limits to immunity, and the cost of vaccine purchase and regulation. Until independent, scientific proof had verified the seriousness of the anthrax problem and the safety and superiority of Pasteur's method of control, the wisest course was to ensure the implementation of proper sanitary regulations.

Regardless of such resistance, Pasteur for a time seemed reconciled to the inoculation of the sheep offered at Junee, always insisting though, on the prior completion of the third, inland, rabbit experiment. But his enthusiasm for the project was waning. By March 1889 he was urging his men to limit vaccinations to a few lots of 10,000, just enough, with controls and public witnesses, to silence the critics. As soon as they had completed the rabbit experiment and vaccinated these few thousand sheep, they were to return to Paris immediately.[40]

Reports in Paris newspapers at the end of March anticipated the Rabbit Commission's unfavourable findings and relieved Pasteur of any lingering sense of obligation.[41] When the commission's official progress report of 3 April made it quite clear that the prize was not going to be awarded to Pasteur, Loir and Germont began packing their bags. Although only two leading scientists on the commission actually insisted that the open country experiment should not be permitted, Pasteur's delegates departed without having carried out the anthrax vaccinations and without having taught anyone the mode of cultivation and inoculation of the vaccine.[42]

Tied from July 1888 to the outcome of the rabbit experiments, the transfer of this product of science had been thwarted. Some aspects of the French disease control system had been assimilated locally, but rabbit politics had proved overwhelming and the determination of parts of the local scientific community to assert their independence of Pasteur's scientific authority had set up apparently unresolvable conflict between the two systems.

Renegotiating transfer

Even so, the demand stimulated by Devlin and his government colleagues now had a life of its own, as a campaign to persuade a still angry Pasteur to supply Australia with vaccine continued across the seas. Still the leading proponents, Devlin and Bruce were now also driven by the representations of many stockowners convinced of the solution to their plight. Past efforts had built up a bigger constituency which developed some political force of its own as anthrax was brought out of the closet and disengaged from the rabbit problem.

Although Devlin had by this time been forced off his property by the ruinous effects of anthrax, other threatened stockowners implored him to persuade Pasteur to establish Australian manufacture. Pasteur eventually agreed, provided the government paid his staff's travelling expenses and the cost of fitting up a laboratory—an estimated total of £1000. Armed with this offer, Devlin and Bruce sought an interview with the new Minister for Mines, Sydney Smith, who was very guarded on the possibility of government funding. A meeting with Parkes confirmed the government's position that it was a matter for private enterprise, and a following letter stated the government's inability to play banker. There was also a limit to the stockowners' readiness to contribute to establishment costs, and less than half the required amount was promised.[43]

Negotiations with Paris were maintained, however, through the continuing efforts of both Germont and Loir, whose own interests proved more compatible than those of Pasteur and more sensitised to the Australian political, economic and geographic environment. Indeed Loir, obviously convinced of the potential demand for the vaccine in Australia, had begun negotiating with Pasteur soon after his return to France, making quite evident his desire for profit in Australia. Persistence was rewarded in October 1889, with Pasteur's transfer to Loir and Germont of full power to negotiate the sale of anthrax vaccine to the Australian colonies.[44]

Now unencumbered by Pasteur's own aspirations and their connection to rabbits, the Frenchmen had a plan to offer the anthrax vaccine monopoly either to the government or a private group under terms which would supply all the Australian colonies for £38,000, or the colony of New South Wales for £28,000. As an inducement to intercolonial co-operation, they suggested that the comprehensive option would cost New South Wales less than £27,000, a marked reduction on Pasteur's previous demands. Such sums would secure only the process, Pasteur having been deterred from sanctioning a manufacturing laboratory because of his mistrust of the 'law of microbes', and the difficulty of defending ideas and interests from 12,000 miles away.[45]

Fortuitously, a cable conveying these plans to Bruce arrived the day before he departed for the Australasian Stock Conference in Melbourne. With Ministerial authority, Bruce submitted the new offer and sought the opinion of the conference as to whether terms arranged with Pasteur should be by lump sum, by royalty per sheep, or by arrangement between owners and Pasteur through a local agent. Although delegates from other colonies were not greatly interested in reopening the subject, the meeting was deftly engineered so as to obtain support from the whole conference, which could be used to justify action on the part of New South Wales.[46]

With ready rebuttals to all objections, the New South Wales delegation first secured a resolution affirming the effectiveness of Pasteur's vaccine. On the financial side, it was clear that the other representatives not only felt unable to advise their governments to contribute to the required £38,000, but thought affected stockowners ought to be left to sort out the whole thing by themselves. Prepared for this, Bruce argued for the compromise of the third option which would allow Pasteur to deal directly with the owners through an agent, but on terms arranged on their behalf by the government. Overcoming dissent, Bruce secured for the record a resolution recommending assistance to Pasteur 'in establishing agencies in all the colonies with a view to those owners who require vaccine obtaining and using it'. Bruce had negotiated the impediment of colonial divergence by offering the other delegates a way of bringing the discussion to some conclusion without apparently committing their governments to anything.

The Department of Mines was now ready to do everything possible to spread the word about anthrax and maximise the demand for the vaccine. In November 1889 it published the Anthrax Board report in booklet form, accompanied by the Junee papers of Stanley and Hamlet and a subsequent paper by Devlin giving a detailed account of the nature of anthrax disease. At the same time Devlin was circulating copies of a letter from Germont to newspapers and to stock boards.[47] On another front, Bruce and his Minister were soon broaching Cabinet with a long minute which revealed Bruce's estimate that anthrax had already cost the colony more than £3 million. It also described the twin evils of current management practices which left land unstocked for four or five months of the year, and risked spreading infection as the sheep travelled on the road. To break the current negotiating stalemate, Sydney Smith urged that stockowner demands for action be accommodated by commissioning an expert to meet Pasteur and canvass all the options.[48]

Events and the demands of the pastoralists were, however, to overtake this strategy. In January 1890 several outbreaks of anthrax were detected, causing the Minister to seek a speedier resolution. With a new minute to Cabinet he won approval to send a cablegram to assess Germont's current position. Guided by Bruce's assessment that the stock conference had only really been opposed to the size and not the principle of the lump sum, the Minister indicated rejection of the £38,000 offer but sought the lowest acceptable figure if an agency arrangement was not agreeable. The response was quick. Germont was not attracted by an agency but he would settle for a lump sum of £25,000 for the Australian rights to manufacture.[49]

No such agreement was made, but by 24 April 1890 Pasteur was writing to Bruce that Loir would return to Australia to establish a vaccine laboratory.[50] Supplied with spores from the Pasteur Institute, Loir would have the full authority of Pasteur for the venture. It was with some relief against the mounting pressure for action, that Sydney Smith announced in early June that the Frenchman was about to arrive.[51]

Forming a French sub-system

On Friday, 6 June 1890, Adrien Loir sailed into Sydney Harbour on board the Messageries Maritimes RMS *Salazia*.[52] The reluctance of Australian governments to secure the vaccine at any price had paid dividends. Not only had the Australia-wide offer been reduced from £100,000 to £25,000, but here was Loir, having incurred the expense of the trip, with no guarantees of recompense, ready to negotiate what he could. He had also brought his own equipment worth £250. The New South Wales government's role as transfer agent had been played without cost and had succeeded in whittling away the French bargaining position.

Loir's negotiations began the day after his arrival, supported by some heavy pastoral lobbying, but the Minister was not inclined to enter into any direct arrangements. For this unpopular outcome, the *Sydney Morning Herald* blamed the head of the Mines Department, while the *Wagga Advertiser* castigated the indifference, bungling and eccentricities

Figure 4.2   The Pasteur Institute of Australia, Rodd Island, 1892 (*La Microbiologie en Australie*, Paris, 1892, Mitchell Library)

of the whole Stock Branch. The chairman of the Pastures and Stock Protection Board at Germanton, John Ross, MP, urged the department to override Loir's excessive price by purchasing the vaccine rights and supplying it for a nominal sum.[53]

Although the government would not countenance any such outlay, it provided the Pasteur venture with several kinds of indirect assistance, beginning with Loir's full use of the facilities on Rodd Island, which had been fitted up for rabbit experiments at a public cost of £2500.[54] Here he was able to live as well as work, saving the expense of accommodation, laboratory and equipment. With a yacht moored at the bottom of the garden, he was installed in some comfort[55] by the end of June, under a six-month agreement which continued until the end of 1893, a period of three and a half years. It was here he was able to entertain Sarah Bernhardt in July 1891.

As Loir began to advertise his wares, the government also provided support in the form of several kinds of public legitimation. When Devlin suggested a trial of the new vaccine, public affirmation of its success was provided by the government's stock inspector for Wagga.[56] When Loir was ready for business, Bruce used all available channels to smooth the way and to stimulate demand. At an agricultural conference convened by the Minister in early July, Bruce again presented a long paper on anthrax, summarising the history of negotiations with Pasteur and indicating his government's willingness to give Loir, as Pasteur's accredited agent, every assistance short of coming between him and the stockowners. To remove any remaining doubts on efficacy, Bruce systematically refuted all the objections made against anthrax vaccination. The department published his paper in full in the booklet detailing conference proceedings.[57] On 9 July, Loir himself was given the opportunity to market his goods to the conference. So delegates could see for themselves the objects described in his scientific descriptions, microscopic photographs were circulated.[58]

In early August, when Loir was sufficiently established and confident for a public unveiling of the process, it was given all the air of an official government occasion. When the steam launch left the floating jetty at Circular Quay on Monday, 4 August 1890, to take a party of inspection to the island, the official group included the Minister, the Under-Secretary for Mines, the Chief Inspector of Stock (Alexander Bruce) and John Lamb, chairman of the Metropolitan Stock Board. It was accompanied by a party of journalists to provide eye-witness accounts for their newspapers.[59]

Loir fed them with many details of this newly transferred technology. After first exhibiting a guinea pig which had died from anthrax, Loir placed a drop of its blood under a microscope so the bacteria could be viewed. He explained the manner in which the microbes were attenuated, and displayed the broth in which they were cultivated. He also described how the 'foundation' for the vaccine, the attenuated spores, would come from France every three months in carefully packed, hermetically sealed glass tubes. On arrival a few drops of the French vaccine were diluted in half a pint of sterilised beef tea and placed in a small chamber at 90°F. Dozens of vessels of beef tea were treated in this way, and after about twelve days the vaccine was ready for use. The visitors were shown bacterial cultures at different stages of growth as Loir described how the final product was sent, again in sealed glass tubes, to Arthur Devlin, who had the contract for inoculation. Particular attention was given to the need and process for sterilising every instrument used in vaccine preparation, including the tubes in which it was cultivated and transmitted from the laboratory.

Loir was also careful to point out, however, that this was a technological system which was still dependent on Paris as its centre. These many physical artefacts of the system had been transferred; a skilled scientist, trained by Pasteur himself, and obviously possessing all this demonstrable knowledge, had also been transferred; yet, as Loir was only too ready to reveal, there were secrets in the cultivation of the vaccine which were not easily mastered; this was why it would at all times be necessary to obtain the supplies of the starting cultures from France. Dependence on the knowledge centre was to be maintained.

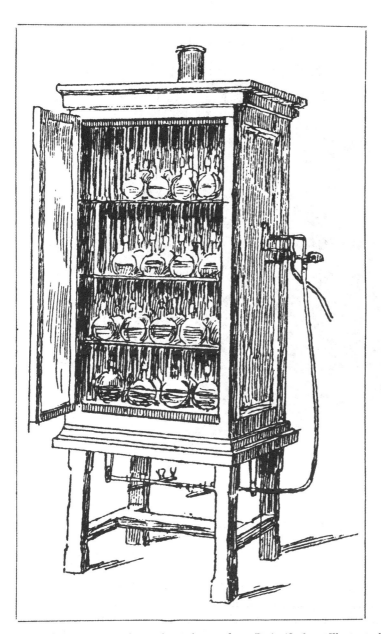

Figure 4.3 Loir's incubator, brought out from Paris (*Sydney Illustrated News*, 21 November 1891)

Figure 4.4   In the laboratory at Rodd Island (*Sydney Illustrated News*, 21 November 1891)

The organisational centre was also in Paris. This new enterprise on Rodd Island, in Sydney Harbour, was called the Pasteur Institute of Australia, and was the first branch of the Pasteur Institute established outside France. The financial ties and responsibilities also went back to Paris. The company backing the venture was announced in Parliament to be the Société de Vulgarisation du Vaccin Pasteur à l'Etranger. Its correct title was actually La Compagnie de Vulgarisation du Vaccin Char-bonneux Pasteur.[60] Directed by M. Sainte-Marie, a businessman interme-diary, this company had been founded to distribute the vaccine to foreign countries and was said to have a capital of 1 million francs.[61] There are conflicting accounts about financial arrangements, but it seems that, having paid a lump sum to the Pasteur Institute for the right to propagate and distribute the spores of the vaccine, the syndicate had a three-year contract to supply Loir at 2d per head, to which he added a further charge to cover his own labour and expenses.[62]

The Australian end of the organisation consisted of Arthur Devlin and his three sons, whom he had been training in inoculation techniques. Devlin himself had learned informally from Germont, during the various trials conducted in 1888. Now he would combine vaccine inoculation with his business as a stock and station agent. Devlin's three-year contract for the vaccinating gave him 1d per head out of the 4d per head to be charged for sheep. For lots over 15,000, charged at 3.3d per head, Devlin would receive 0.3d.[63] By November 1890 a Victorian agent, a local vet-erinary surgeon, was also ready to begin inoculations.[64]

With an operational organisation and a marketing pamphlet in circu-lation,[65] Loir expressed great confidence to his mother. In a letter of 21 November, he indicated that M. Sainte-Marie would receive in the vicinity of 25,000 francs for the vaccinations already made.[66] It was as well that stockowners did not know, the cost of vaccination being somewhat con-tentious. For this reason, stock boards were still pressing the government to take over the business.[67] Loir was happy to add fuel to the argument, informing the Minister and the public that Hungarian stockowners paid only 2d per head because a government-run laboratory made the vaccine.[68] The government remained unmoved, yet it had already done much. Through its own actions and those of its officers, it had played a large part in bringing this scientific product to Australia, and in gener-ating the publicity and demand needed to make Loir's private venture a success.

And successful it appeared likely to be. During July 1890 Arthur Devlin and sons carried out the first commercial vaccination at Brocklesby station in the Corowa district, where losses had often been as high as 20 per cent.[69] Other neighbours soon followed suit. In fact, during the 1890 season almost 200,000 sheep were vaccinated in New South Wales.[70] Through a combination of efforts, spurred on by stockowners' demands, anthrax vaccination had at last been transferred to Australia and the process for its manufacture and distribution established.

CHAPTER FIVE

From Paris to Narrandera

The Australian pastoral industry had already demonstrated the capacity to assimilate some of the precepts of Pasteur vaccination into its own sub-system of disease control, thanks to the efforts of some key people in both its public and private sectors. Now that the French technological system had established its own branch in Australia, with formalised links into the Australian system, the diffusion of its product depended on acceptance by a continually widening clientele. It was an acceptance not easily won.

A slight delay in diffusion

The use of Pasteur's vaccine over the next eight years was hesitant, erratic and soon on the decline, never again reaching the levels of the first season. The reasons reflected factors of supply, of demand and of environment.

The first set of obstacles facing the Pasteur vaccine was clearly related to its technical performance under Australian field conditions. In the 1890 season, the losses of vaccinated flocks, at around 5 per cent, were considerably higher than the figures of less than 1 per cent reported for France and other European countries.[1]

In his analysis of this 'partial failure', Bruce as Chief Inspector of Stock ruled out any fault at the laboratory level because of the stringent testing before the product's despatch, but he did concede that environmental factors could have played a part. Loss of potency from exposure to heat or light was possible either during the long coach journey from Sydney, or during the actual operation, especially under summer conditions. It was also possible that two doses of vaccine might not impart sufficient immunity to protect against spore-laden pastures infected by diseased carcasses left to rot. To allow for this, Loir was prepared to revaccinate whenever vaccination was unsuccessful.

Figure 5.1   A contemporary sketch of inoculation with Pasteur's vaccine
(*Sydney Mail*, 25 October 1890)

Loir, Devlin and most of the owners laid the major blame for losses on operating in hot weather when there was also the danger of vaccinating sheep already incubating the disease. For this reason Loir decided to limit vaccinations to the milder period between the end of May and the middle of September. With this concession to Australian conditions, the results for 1891 were much improved, with total losses reported at less than 0.5 per cent.[2] But reports in 1893 of outbreaks among sheep in *inoculated* districts[3] suggested further failures, and Arthur Devlin himself conferred with Loir about the 'occasional' high mortality rate.[4]

There were also economic frictions between the Pasteur vaccine and the Australian environment. In contrast to the European way where small numbers of sheep were confined to restricted pastures, Australian flocks were pastured in large numbers over wide areas. Mustering sheep for inoculation was therefore time-consuming and expensive, particularly given the 'scrubby' nature of much of the country at the time. Two musterings a fortnight apart presented a far greater burden than in Europe. There was also a greater risk, for the chance was higher that some animals might be missed during the first inoculation, in which case the second, stronger dose would probably kill them. Added to this was the high cost of labour for the musters, amounting to something like 25 per cent of the total cost of vaccination,[5] more where a third or fourth inoculation was necessary, as was often the case.

Rumblings of discontent continued about the price of the vaccine. No one would pay 4d per head, or even the 3.3d for large herds, unless thoroughly assured that losses from anthrax would exceed the cost of vaccination. Even where yearly losses amounted to many times more than this, many owners resisted paying out ready cash in advance for protection.[6] Compounding their resistance was the limit to immunity and Pasteur's recommendation of annual vaccination. Thomas Hammond said: 'It pays me better to send the sheep up to the mountains than to have them vaccinated.'[7]

Though Adrien Loir congratulated himself on negotiating a reduction in the vaccine cost during a return to France in the first half of 1892, enabling him to advertise a new basic charge of 2d per head for sheep and 4d for cattle,[8] Table 5.1 suggests that it had no great effect on demand. Sheep numbers picked up in 1893, although it is notable that the number of lots did not. Though small stockowners often combined their herds to take advantage of the sliding scale of charges, there was clearly no acceleration in the use of the product. It seems price was not the only barrier.

Table 5.1 Use of Pasteur vaccine, New South Wales, 1890–1898

Year	Sheep vaccinated	No. of lots
1890	198,098	—
1891	98,261	12
1892	70,029	16
1893	119,380	16
1894	174,760	10
1895	36,400	4
1896	7,320	3
1897	16,017	—
1898	61,270	22

Source: Annual Reports of the Stock and Brands Branch of the NSW Department of Mines

There was, of course, the general economic state of the industry. Wool prices had been falling steadily over the previous decade. The average value per bale dropped from over £20 in 1880 to £12 in 1892, accompanied by a similar fall in the average value of wool per sheep.[9] At the best of times Australia's less intensive pastoral industry experienced vaccination costs as a relatively greater burden than the European industry. In these less happy times, the bite was even more savage. To make matters worse, the 1890s were bringing a period of general economic depression. As a land boom gone bust deepened in late 1891 and became widespread financial crisis in 1893, one of the specific effects on the pastoral industry was the withdrawal of finance lent so freely during the 1880s.[10] No wonder there was resistance to cash outlays to prevent losses that might not even eventuate.

Meanwhile, a continuing element of opposition to the practice of anthrax vaccination preyed on the fears and suspicions of many unde-cided stockowners, especially when it came from scientific authorities like Oscar Katz[11] and Dr William Camac Wilkinson. In Parliament, John Stewart's regular niggling in the Legislative Council found its counterpart in Dr Andrew Ross's efforts in the Legislative Assembly. While Stewart was keen to draw attention to the fact that deaths were occurring as a result of anthrax vaccination,[12] Ross seemed obsessed with the idea that vaccination caused cancer and tuberculosis. He was also concerned that stock which had undergone repeated inoculations of 'virus' would com-municate these diseases to humans through their meat or milk, and seemed resentful of priority given to animal diseases over human disease.[13]

The intensity of parliamentary questioning peaked during the latter part of 1892 as Loir returned from his trip to Paris and resumed residence at Rodd Island. Another dimension was now added to the opposition to vaccination, as residents of surrounding areas expressed concern about this use of the island. There were two aspects to their indignation. One was fear of 'the microbes'. The other was an assertion of their rights to reclaim the island's use as a public recreation area, as declared in 1879.[14] Wilkinson added his voice to the general concern, drawing attention to Loir's unsuccessful experiments at Tenterfield with an inoculation for blackleg, and to the government's right to manufacture vaccine without payment for a non-existent secret.[15]

The local touch

Against this background, some reticence among stockowners is perhaps not surprising. Yet from 1894, the diffusion of anthrax vaccination accel-erated rapidly as an alternative product became commercially available. The man responsible was John Alexander Gunn, an insider of the indus-try, the manager of a Goldsbrough Mort property in the Narrandera district.

John Gunn was young for such responsibility. He had left his home in Buninyong, Victoria at the age of eighteen, to become a jackeroo at the South Yalgogrin station near Wagga Wagga in New South Wales. Soon he was overseer; by 1886, he was still only twenty-six but was manager of the whole station. For Gunn was true to his Scottish heritage, alert, industrious and energetic. His Goldsbrough Mort superiors described him as a 'first class man', and a 'very careful and efficient manager'.[16] Others described him as a 'leader', and 'big-hearted'.[17] When Sidney and Beatrice Webb visited in 1898 they thought him 'a thorough-going individualist, hard-working and upright', with a keen 'speculative intelligence' and commercial instinct.[18]

He was also astute and well-informed on the happenings within his industry. And his talents were well-rounded; his experience with Aboriginal people around the Narrandera–Wagga area had already been transformed into a novel called *The Canargo Blacks*; and he had improvised, improved and patented the first mechanical poison-bait layer. Called Gunn's Patent Pollard Distributor, it was manufactured by the local firm of Lassetter and Co and was soon in general use. But Gunn was no mere tinkerer; he had qualified as a surveyor, even made his own theodolite; then he purchased a microscope to support his interest in the study of insect and plant life.[19]

Gunn, too, had known the anguish of dealing with anthrax. In October 1888 he had watched the Junee demonstrations of Pasteur's vaccine 'very minutely', returning home to begin his own microscopic examinations of the blood of diseased sheep on South Yalgogrin. He was soon convinced that the mystery disease which had demolished his own flocks was indeed anthrax. He was ready to embrace this wonder vaccine from France.

Gunn used the anthrax vaccine in 1890, the first summer of its commercial life in Australia. He was pleased to have it, but frequent difficulties led his questioning mind to experiment. With characteristic tenacity, eventually he produced his own vaccine (a process described in detail in chapter 6). In the summer of 1893-94 Gunn vaccinated Yalgogrin flocks and those of his neighbours, with remarkable success. The following summer he was swamped with orders from more distant quarters and began large-scale production. Then in April 1895 Gunn joined up with John McGarvie Smith to form the firm of McGarvie Smith and Gunn. Smith also had Scottish origins, and a bent for the industrious and studious life. But he was a city man, originally trained as a watchmaker and jeweller, then later as a chemist, metallurgist and finally bacteriologist.[20] It was a partnership which would change the course of Australian vaccination.

Acceptance of the McGarvie Smith and Gunn vaccine continued to rise. In tandem ran the declining use of the Pasteur vaccine, as the Australian product assumed dominance and virtually displaced the Pasteur product from the market. Loir had left Australia early in 1893, to be replaced by another Pasteur 'savant', Dr Louis Momont, and then Dr

Figure 5.2   John Gunn with his two sisters (R. H. Webster, *Bygoo and Beyond*, Ardlethan, 1955)

Emile Rougier. But by 1897 the Pasteur business had so run down that Arthur Devlin withdrew to become a stock inspector[21] and in early 1898 Rougier sold the agency in Sydney to Frederick Zimmerman, who operated under the name of the Pasteur Anthrax Vaccine Laboratory.[22] By 1903 it was no longer listed in the Sydney municipal directory.

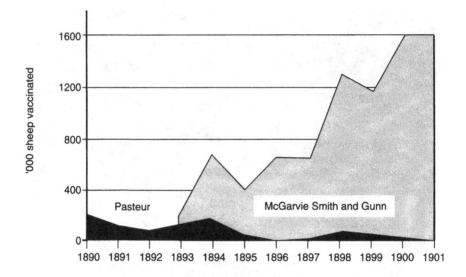

Figure 5.3   Comparative use of the competing vaccines in New South Wales, 1890–1901
Source: Annual Reports of the Stock and Brands Branch of the NSW Department of Mines

All available data imply that it was the Australian product that hastened anthrax vaccination from 1894. That year brought an increased general demand, and sales of the Pasteur vaccine also rose. For it was a year of mounting concern about the spread of anthrax, and in July 1894 a special meeting of representatives of the stock boards of the colony was well attended.[23] Several papers stressed the seriousness of the situation, and a motion urged the government to introduce legislation making vaccination compulsory. Only two people spoke against the motion, which was carried overwhelmingly.

Yet this heightened awareness does not explain why stockowners swung so strongly to the new Australian vaccine. The accepted wisdom, attested to by several leading pastoralists,[24] said that the Australian vaccine was simply infinitely superior in performance. One pastoralist insisted that Pasteur's vaccine 'was a failure and it killed sheep'.[25] Another, writing in February 1896 to confirm the 'thorough success' of Gunn's vaccination, confessed he 'was a bit doubtful of the results at the start, but you cannot blame me. My experience of Pasteur's people was enough to shake anyone's faith in vaccination.'[26]

As Gunn told the story, it was the recurring trouble with the Pasteur vaccine and his desire for *fresh* material that led him to begin experimenting himself,[27] a version supported by the independent testimony of a Yalgogrin employee and of watchful neighbours.[28] The changing tenor of his own views is also evident in his public statements. In letters and articles sent to the press from October 1891,[29] Gunn was an early advocate and defender of anthrax vaccination and Pasteur's vaccine. Yet behind his protestations of allegiance, there were signs of difficulty. Around June 1891, he reported to the local Land Board that the 1890 vaccinations on South Yalgogrin had achieved 80–90 per cent protection only after four inoculations.[30] By December 1892, Gunn was still ready to defend Pasteur's vaccine against the criticisms of Dr Wilkinson,[31] but a more general note of qualification was evident as he admitted loss of potency to be more than a teething problem of the first season. Nevertheless, he insisted that where losses were large, vaccination was infinitely preferable to allowing the disease to rage unchecked.[32]

By May 1893, Gunn's praise for Pasteur's vaccine was even more tentative. Theoretically, he said, the process was perfect and in Pasteur's own hands, on the small numbers of old world flocks, it had in general proved itself so in practice. But Australia imposed greater strains on the vaccine. Admitting 'the trouble and loss' caused by variable potency, he looked forward to the future availability of a product which would be much superior in performance, one which 'while conferring the same protection as given by Pasteur's vaccine at its best, will be of uniform and unvarying strength, and can be relied on for any length of time'.[33]

The enigmatic McGarvie Smith would later attribute the failures of the Pasteur vaccine to the fact that Loir's product 'was teeming with motile micro-organisms', while that of his successor, Momont, 'was no better . . . his failing being he was fond of the wine'.[34] But the comparison of the mortality rates associated with the two vaccines in Table 5.2 suggests something more complex.

There is indeed a puzzle here. Pasteur's vaccine did seem to be performing better. Associated deaths for 1891, 1892 and 1893 hovered around 0.4 per cent, a result quite comparable with the European figures, hardly justifying the swing to Gunn in 1894. Perhaps mortalities reported to the Stock Branch were understated, but it is more likely that when the Pasteur vaccine failed in Australia, it did so in a big way, on particular lots, with a significant impact on perceptions. The year 1891 saw one lot with a high death rate of 4.1 per cent, and in 1892 another had a failure rate of 3.4 per cent. Then 1894 brought a couple of bad failures, in one case 2000 out of a flock of 14,000. Such losses early in the season may have influenced stockowners to try the alternative. Yet, for 1894, we see a similar pattern of individual failures for Gunn's vaccine; 14.6 per cent were lost in one case, 10 per cent in another; in several other cases, losses lay between 1 and 3 per cent. Still the stockowners voted for Gunn's vaccine again in 1895. Why were they more forgiving of Gunn's vaccine than Pasteur's?

Table 5.2 Deaths associated with anthrax vaccination, New South Wales, 1890-98

Year	Pasteur Vaccine			Australian Vaccine		
	Sheep vaccinated	% deaths	Lots affected*	Sheep vaccinated	% deaths	Lots affected*
1890	198,098	approx. 5	—			
1891	98,261	0.37	2/12			
1892	70,029	0.44	4/16			
1893	119,380	0.38	3/16	50,000	0.05	
1894	174,760	2.0ᵃ	3/18	441,646	<1.5	28/41
1895	36,400	16.5	1/4	310,298	0.7	20/45
1896	7,320	>5ᵇ	2/3	643,736	0.12	20/74
1897	16,017	0.17	3/17	627,337	0.10-0.13	12/103
1898	61,270	no details		1,166,106	no details	

Source: Annual Reports of the Stock and Brands Branch of the NSW Department of Mines
* Lots affected expressed as a ratio of total number of lots.
a Losses were 1.3% for half of the vaccinated stock, plus one big failure, together making an overall estimate of 2%.
b Losses unspecified, but described as very large.

By 1894 it seems that Pasteur's product had already built up a reputation for unreliability through these occasional dramatic failures. In 1895 the loss of 6000 from a flock of 30,000 would have done nothing to restore confidence. To nervous stockowners the well-publicised success of the Gunn vaccine over the summer of 1893-94 must have been welcome news. Gunn's own descriptions of his experiments in the *Australasian*[35] were followed in March 1894 by items in the *Narrandera Argus* and the *Australasian Pastoralists' Review* rejoicing in the success of the 4000 inoculations so far carried out, reporting how on one station, 'a previous vaccination by another specific failed to counteract the effects of the disease'.[36] By June it was reported that Gunn had inoculated about 50,000 sheep 'and so successful has been the result, that the deaths have been only about 25 head all told'.[37]

This was impressive news. But perhaps there were other incentives for chosing the Australian vaccine. There is the suggestion of uncertain supplies of the Pasteur vaccine in Gunn's reference to a letter received from Devlin in 1893: 'he states that he like myself can get neither answers nor vaccine from Pasteur's people'.[38] There was also the pull of a substantial price differential, with Gunn's first commercial season ranging from 2.5d per head down to 1.5d,[39] against the Pasteur scale of 3d to 2.3d. And, with the 1894–95 season on its way, where better for Gunn to distribute a new advertising leaflet than at the July stock board conference on anthrax. The Pasteur people also saw this as a significant event in the marketing of their vaccine, for Dr Momont was present. But Gunn had the advantage in these circles. He was by nature blessed with a commanding physique and personality; now he was also a person of some standing in the pastoral community. As a member of the Narrandera Stock Board, he was present as the delegate for the Narrandera Sheep District and spoke in favour of the motion for compulsory vaccination.[40] We know he had secured written testimonials in time for the meeting from those who had used his vaccine,[41] so Gunn was bound to have taken this most fortunately timed opportunity to put his case. Some of the conference participants later appeared as customers in the books of McGarvie Smith and Gunn.[42]

As manager of Yalgogrin, Gunn was also part of the Goldsbrough Mort network of pastoral properties. And his authority there was increased by his 1895 appointment as superintendent of all stations in southern New South Wales. On at least one of these stations, the reduced incidence of anthrax by 1896 was officially attributed 'in great measure to the fact that all the sheep before being sent to Retreat had been inoculated'.[43] The evidence suggests, then, that Gunn's own position and credibility within the pastoral community were probably a factor in the take-off of the vaccine in 1894. Such standing was extended after December 1895 when he was given the authority to represent Goldsbrough Mort at the Narrandera Land Board.[44]

So a range of factors gave Gunn a chance in pitting his home-grown product against the authority of Pasteur. Yet these alone could surely not sustain and increase diffusion without product performance. So what gave the Australian vaccine the edge?

Adapting the product

When Gunn first produced a commercial vaccine it was modelled on the Pasteur product, although he insisted that his *method* of production was unique.[45] It too was a broth culture, somewhat vulnerable to temperature, and had to be used within five days of production. Of the half million sheep vaccinated by Gunn in 1894, the failures occurred 'chiefly in districts remote from the laboratory, and where the time conveying the vaccine to the work was very great'.[46] Gunn had the advantage of being located in the heart of anthrax country, and with his team of vaccinators, and a network of contacts to provide fresh horses, was usually able to distribute within the time limit. Still, Gunn was working on trying to eliminate this restriction.[47]

Collaboration with McGarvie Smith seemed to offer the solution. Smith brought to the partnership a method which improved keeping quality. The answer lay in the antiseptic properties of glycerine, which kept bacterial contamination at bay in a glycerine-water emulsion of the vaccine culture. By July 1895 it could be claimed that the deterioration problem had been solved.[48] But in fact a more effective solution was on the way.

Another defect Gunn had been addressing was the necessity for two inoculations a fortnight apart. As a pastoralist he was only too well aware of the cost, risk and inconvenience this imposed under Australian conditions. But it was only with much persistence, in conjunction with McGarvie Smith, that a resolution in the form of a *single-dose* vaccine was developed. Hand-in-hand with this innovation went the preparation of a spore-based vaccine.

The single dose was managed by finding a level of attenuation that was strong enough to induce adequate immunity but not so strong as to kill the inoculated animal. It was something which Pasteur insisted could not be safely done.[49] It was certainly inherently more risky than developing immunity in two steps: at a strength which was just sub-lethal for sheep, the attenuated culture had to be exactly right. This was controlled by a stringent method of testing which rejected any batch which did not completely measure up. That level of attenuation was then maintained in transit and storage by the resistant spores. Introduced in 1897, this marked a major advance for the stockowners of Australia and secured the anthrax vaccine market for the firm of McGarvie Smith and Gunn.

The choice before Australian stockowners was not, therefore, between two like but competing products. During the troubled 1890s pastoralists were in fact offered four separate innovations, as a result of research developments which will be described in the next chapter. To summarise:

- The first was Loir's two-dose broth vaccine, introduced in 1890. It was prepared in Sydney from attenuated spores sent from Paris but was not itself a spore vaccine.
- The second was Gunn's two-dose broth vaccine, introduced in late 1893. Produced in Narrandera, it was apparently more reliable than the Pasteur vaccine.

- The third innovation came with the McGarvie Smith and Gunn vaccine, a glycerine-water emulsion with improved keeping qualities, introduced in 1895.
- The fourth actually combined two interdependent innovations in the McGarvie Smith and Gunn single-dose spore vaccine, introduced in 1897.

To attempt to fit the single, S-shaped line of the usual diffusion graph to Figure 5.3 would therefore be somewhat misleading. It is not simply the story of increasing acceptance of a single unchanging product. Rather, it represents a dynamic process of incremental adaptation to the specific climatic, geographic and economic conditions existing in Australia. This becomes even clearer if we examine the way anthrax vaccination spread.

Patterns of diffusion

An insight into how these vaccine improvements accelerated its diffusion can be gained from delving into the vaccination returns of the firm of McGarvie Smith and Gunn. Three features are of note. Firstly, the typical approximate S-shape which appears in Figure 5.4 indicates that adoption of the McGarvie Smith and Gunn vaccine was not immediate amongst all its potential users, even though its performance was allegedly superior to that of the Pasteur vaccine. Many stockowners took some time to be convinced of the need for vaccination, the plateau in the curve being reached about 1901, several years after the establishment of the firm of McGarvie Smith and Gunn. There was, however, a definite step up in the rate of adoption in 1897–98, corresponding with the introduction of the single-dose vaccine (Figure 5.3). This was obviously an attractive feature. Thirdly, analysis of the number of vaccinating properties indicates that the diffusion of anthrax vaccination was due to a widening of its acceptance among an increasing number of stockowners, rather than a deepening of its use by the same pool of pastoralists. The decline in the average number of sheep being vaccinated on each property confirms this trend (Figure 5.4).

The 'neighbourhood effect'

Spatial factors affect information flows and personal communication and were clearly evident in the initial stages of anthrax vaccination. The early geographical concentration of users around the Narrandera area, as

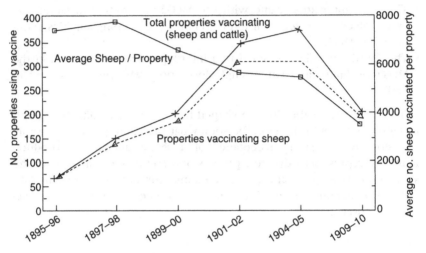

Figure 5.4   Use of McGarvie Smith and Gunn vaccine on properties in New South Wales and Victoria, 1895-96 to 1909-10
Source: Vaccination returns, McGarvie Smith and Gunn

evident in Figure 5.5, reflects this region's far greater exposure to vaccination: the informal then formal diagnosis of the identity of Cumberland disease on Devlin's property, and Devlin's active campaign among surrounding pastoralists and sheep boards, were followed by the public trial of the vaccine at nearby Junee. The latter was a big event for the region, bringing publicity to the cause of anthrax vacination, visible proof of its effectiveness and some familiarity with its procedure, techniques and theory. It was just this sort of exposure which first convinced Gunn that his problem was anthrax and the Pasteur vaccine was the solution. His sheep were vaccinated from the first season.

So when Gunn began experimenting, other property-owners in the area were aware of his activities and of the concept and practice of anthrax vaccination. Indeed, it was within this general area of the colony that Devlin had begun using the Pasteur vaccine. While Gunn could take advantage of the relatively high level of local knowledge, this could also be a disadvantage if perceptions of failure of the Pasteur vaccine resulted in a lasting bias against the practice of vaccination. But Gunn's personal contacts and standing in the community were helpful in persuading people to try his alternative. Once this step had been taken, geographical proximity to the source provided another advantage, since Gunn's vaccine could be delivered quickly, before deteriorating, ensuring superior uniformity of performance over the Pasteur product.

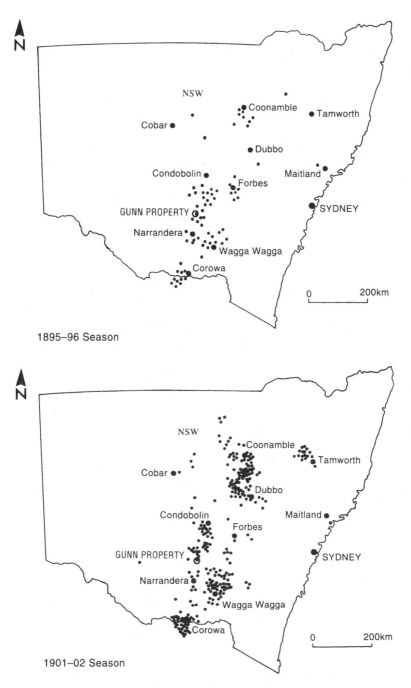

Figure 5.5 Spatial diffusion: properties vaccinating with McGarvie Smith and Gunn vaccine

Sources: Vaccination returns, McGarvie Smith and Gunn; *Australian Pastoral Directory*; H. E. C. Robinson's Pastoral Map of New South Wales

There were good reasons, then, why the primary cluster of users were located in the areas around Narrandera; in this case the 'neighbourhood effect' described by Hagerstrand[50] had both information *and* performance consequences. The significance of the improvement in keeping quality, introduced in the 1895–96 season, thus lay in its extension of the safe geographical range of distribution of the vaccine. As the *Australasian Pastoralists' Review* declared, 'the whole problem of the complete protection of sheep from anthrax at any distance from the laboratory and in any temperature is now solved'.[51]

Such pronouncements in a major pastoral publication were no doubt effective in opening the market further afield. Despite the initial lag in the areas further away from Narrandera, they soon caught up. Figure 5.4 indicates the general plateauing of diffusion in the early years of the new century. A similar trend was evident in all the major vaccinating districts. By the turn of the century, longer-range information flows were taking effect. These included the regular items appearing in pastoral journals like the *Australasian Pastoralists' Review* and newspapers like the *Australasian*, reporting the success of the Australian vaccine. Supplementing these were regular advertisements placed in pastoral publications by McGarvie Smith and Gunn, and leaflets for distribution each season, showing the current scale of prices. As word and confidence spread, new clusters of users were started.

Assessing and managing risk

New clusters carried vaccination further, yet it is notable that anthrax vaccination remained essentially confined to the central strip of New South Wales and Victoria, a geographical limit which does not simply define the major sheep areas of the colonies. Certainly some of the areas using the vaccine were among the districts which carried the most sheep, but not all, and several other prominent sheep areas did not take up vaccination.

It could be that at these spatial boundaries we see the fading out of the 'neighbourhood effect' which emanated from the South Yalgogrin property in Narrandera. However, the more likely explanation lies in the coincidence of the vaccinating regions with what has come to be known as the 'anthrax belt'. Before the introduction of anthrax vaccine, Cumberland disease had been identified in several areas in New South Wales, including Grafton, Tamworth, Nyngan, Dubbo, Mudgee, Young, Albury, Narrabri, Bathurst, Forbes and Goulburn, most of which were in the central division of the colony. It was also stated in 1888 that anthrax existed in the areas around the Gwydir, Namoi, Castlereagh, Macquarie, Bogan, Lachlan, Murrumbidgee and Murray Rivers. This corresponds strongly with the areas in which anthrax vaccination became established. It also corresponds with the area in which future outbreaks were recorded in the period 1935-1951.[52]

What this tells us is that the use of the vaccine was very closely linked with a high level of perceived and demonstrated threat of anthrax infection, as when Donald Fletcher of Balagula judged from past experience that there was 'very little room to doubt that I would have lost at least 50% had I had uninoculated sheep in that paddock all summer'.[53] Informed fear drove vaccination; it did not become a standard precaution. Anthrax vaccination was not an innovation which was in itself capable of generating profit, only of reducing losses. It was a cost to be deducted from profit and only those who had personally experienced the devastation of anthrax were prepared to put risk before cash. This was despite the infectious nature of the disease and the fact that it could be transmitted by spores in the ground and spread by travelling stock, the factors which led pastoralists in several areas to press the government for compulsory vaccination.[54]

But vaccination did not become compulsory; in fact the practice began to decline toward the end of the first decade of the twentieth century as the vaccine's effect in controlling the disease reduced the perception of threat.[55] The link between risk and use meant that, as a product, anthrax vaccine had inherent limitations for market expansion; the more effective it was, the more it eliminated the motivation for its use.

There was also the fact that stockowners were learning to manage the vaccination of their flocks to obtain the best effect within their budget constraints. This was a skill which developed with familiarity with the technology. Even within the anthrax belt, vaccination was never near saturation levels. The five major vaccinating districts of Narrandera, Wagga, Condobolin, Coonamble and Corowa were inoculating about 20 to 30 per cent of sheep at the peak of vaccination.

Nor was saturation vaccination ever likely. Even though Pasteur had recommended annual vaccination to guarantee immunity, Loir predicted that only a small percentage of vaccinated stock would be vulnerable to infection even two years later. He thought ordinary flock sheep would need to be vaccinated only once every three years. Only the most valuable sheep justified annual vaccination.[56] In fact, stockowners grew increasingly comfortable with the practice of limiting vaccination to new stock and to flocks where infection was detected.[57] The result was a fall in the vaccination rate on individual properties to a level that was considerably lower in 1901-02, when vaccination was reaching its peak, than it had been a few years earlier (Figure 5.4). Stockowners were getting better at optimising benefits and costs.

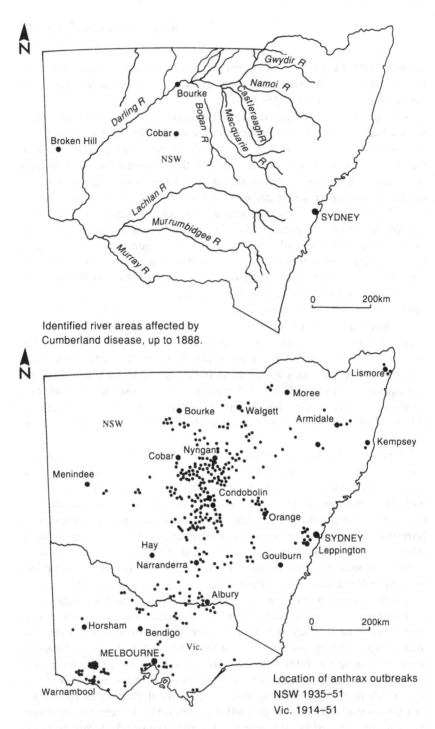

Figure 5.6   The New South Wales anthrax belt
Source: H. R. Seddon, *Diseases of Domestic Animals in Australia*, vol. 1, part
5, Canberra, 1955

Shifting benefits

This decline in the vaccination rates on individual properties is also related to another significant factor in the nature of the diffusion process, namely the size of the vaccinating properties. We have already seen that the leaders in the use of the McGarvie Smith and Gunn vaccine were mainly located relatively close to the source. The leaders also tended to be large property-owners with large flocks. In New South Wales in 1895–96, only 6.9 per cent of all properties carried more than 10,000 sheep, compared with 57.6 per cent of vaccinating properties. Likewise, 66.2 per cent of the properties in New South Wales carried less than 1000 sheep at this time, and none of these was vaccinating for anthrax.[58]

But this early skew towards large properties was gradually modified as increasing numbers of smaller property-owners adopted vaccination. By the 1904–05 season, the picture had changed quite markedly. Now only 20.6 per cent of vaccinating properties held more than 10,000 sheep, compared with the 4.4 per cent for all of New South Wales. And at the other end of the scale, 14.7 per cent of those vaccinating carried less than 1000 sheep, and 62.6 per cent carried 5000 or less. The smaller stockowners were increasingly responsible for swelling the ranks of anthrax vaccine users.

There are logical reasons to expect the large property-owners to take the lead. Advantages in access to finance combined with readier access to information. Managers or owners of large properties were more likely to be involved in pastoral affairs, to be elected to sheep boards, to represent their district at meetings and generally have greater exposure to channels where new developments were discussed. Reinforcing these advantages were lower per capita vaccination costs because of the discounts provided for large flocks. However, the evidence suggests that, prejudices and fears aside, owners were tending to take a cost-benefit, or risk-benefit, approach to the matter of anthrax vaccination, and the cost-benefit ratio was becoming more and more favourable to the small owner.

In the early years, before he began commercial production of his own vaccine, Gunn himself was advising that where stock losses were likely to be 5 per cent or greater, it was economically rational to vaccinate.[59] This calculation was based on a vaccine cost of 3d per head plus an inoculation cost of 0.5d per head. Gunn's own vaccine was initially offered at a total cost of 2.5d to 1.5d, depending on the size of the flock. The McGarvie Smith and Gunn partnership opened with prices ranging from 2.25d per head down to 1.5d. By

1900 the scale ranged from 2d to 1.5d, with special discounts on very large lots. On top of this, the single-dose vaccine had reduced the labour cost of inoculation by at least half. By 1902, the general rate for the vaccine was 1.5d per head, with discounts for lots over 20,000.[60]

What was happening then, was that the vaccine cost, in particular for small herds, was reducing all the time. As the cost of vaccination fell, it became less and less rational to risk losses by not vaccinating. There is no evidence of an increasing level of risk of anthrax outbreaks toward the end of the century when the rate of vaccination was accelerating. Instead, increasingly, the small owner was able to cross the threshold where cost no longer outweighed the assessed risk.[61]

As more recent studies of diffusion have shown,[62] potential adopters of a new technology cannot be perceived as a uniform population with a uniform set of decision-making criteria. The boundary of acceptability of a new technology moves with changes in the technology and with the widening circle of potential users. And in Australia at the end of the nineteenth century the boundary was being shifted through the adaptive efforts of two Australians.

Of course other factors were operating, such as the gradual recovery from the depression toward the end of the 1890s. But in the cost-benefit approach to vaccination being taken by stockowners, it seems that the McGarvie Smith and Gunn vaccine was generally rated as having those qualities which made it worthwhile to outlay cash for vaccination, while the Pasteur vaccine did not.

The Pasteur vaccine, and the activity and information associated with its demonstration and use, had clearly been of advantage in conditioning the initial adoption of the Australian vaccines. Equally clearly, Gunn's and McGarvie Smith's modifications to the product affected the cost-benefit ratio in several important ways. Firstly, the consistent reliability resulting from their method of production reduced the risks involved in the actual use of the vaccine. Secondly, it cut the cost of the vaccine to users, especially the small stockowners who could not benefit from the discounts previously available to large properties. Thirdly, by making a single dose safe and effective and thereby eliminating subsequent inoculations, it not only removed another risk from the procedure but, very importantly in the Australian context, dramatically cut the labour cost associated with vaccination.

It was the Australian investigators, John Alexander Gunn and John McGarvie Smith, who responded to the unique climatic, geographic and economic circumstances of the Australian pastoral industry and delivered these benefits over and above what Pasteur's vaccine had to offer. Their incremental experimentation and refinement made the diffusion of anthrax vaccination a reality in Australia. Adaptation was the key to diffusion, and the Australian system had shown its ability to generate significant technological change.

CHAPTER SIX

From foreign to domestic capability

A spreading French system

The French technological system which evolved under the direction of Pasteur and took institutional form under the guise of the Pasteur Institute in 1888, continued to grow and establish its roots around the world. Organised around the fundamental principles and techniques of attenuation and vaccination, it became a centre of research on virulent and contagious diseases, a teaching centre and a manufacturing centre for animal vaccines for international distribution. Students came from all over the world to learn at the source, some staying, some returning home to spread their new knowledge, sometimes even starting up new institutes.

One of the most effective means for spreading the new fund of knowledge building up under the impetus of Pasteur's colleagues and protégés was the official periodical of the institute, *Les Annales de l'Institut Pasteur*. Begun by Emile Duclaux in 1887, this carried the objects, methods, equipment and results of the Pasteurian system to all quarters of the scientific world.[1] Within a decade of the opening of the Pasteur Institute, two branches were operating inside France and nine institutes were operating overseas. This would eventually grow to thirty.[2]

The transfer of the manufacture of anthrax vaccine to Australia through the Pasteur Institute of Australia represented part of the process. As the first branch of the Pasteur Institute outside France, it required the establishment of organisational, physical and institutional commitments. Some intellectual allegiance to the Pasteur tradition had already begun in 1888, with the 'scientific proof' at Junee and all the associated debates. Those who were determined to see the vaccine available in Australia fought to make a connection in the minds of pastoralists between the mysterious processes in a laboratory on the other side of the world and the alarming events occurring in their fields and account books.

Through meetings, pastoral publications and even daily newspapers, the message was spread. Scientific comprehension might not have been high, but the pastoral industry and its administrative caretakers were introduced to the terminology, the concepts and the experimental proof of the proposition that the answer to their problems lay in the product of a scientific laboratory driven by the scientific knowledge and techniques of the Pasteur tradition.

Loir's personal efforts helped to spread the word through the scientific community. As a member of the Royal Society of New South Wales, he presented papers on anthrax vaccination and rabbit disease.[3] Joint experimental work with Stanley included investigation of the kangaroo's susceptibility to anthrax, and the rabbit's susceptibility to various infectious diseases.[4] Collaboration with Alexander Bruce resulted in a paper published in *Les Annales de l'Institut Pasteur*.[5] Loir was also conducting other experimental work on local stock disease, including pleuro-pneumonia, blackleg and tuberculosis, all of which heightened the perception that the Pasteur approach to disease control was relevant and important to Australia.

The transfer of scientific capacity

With acceptance of the Pasteur view of the world came enthusiasm to emulate its scientific character in the form of an Australasian Stock Institute. The proposals emerged from the 1889 Intercolonial Stock Conference and arose from first-hand experience of what Pasteur's representatives could achieve.[6] Yet it was clear that the intent was not to reinforce scientific dependence on Paris. Rather, the desire was to transfer all the elements necessary to develop a local capacity for science. Though headed by a fully qualified European scientist, this proposed new institute was a creature of the Australian system: it would wage war on Australian stock diseases through local scientific research, teach microbiology to Australian students, and be funded and run by a board of management representing each colony. The board would direct the work of the scientist and the conditions of tutelage.

It was agreed that New South Wales was the appropriate location and the 'hearty' endorsement of the 1889 resolution by the New South Wales contingent was followed in August 1890 with submission of a detailed report from Lamb, Bruce and Stanley to the appropriate Under-Secretary.[7] By this time Adrien Loir had returned to Sydney to manufacture anthrax vaccine and could give first-hand and up-to-date details of how the Pasteur Institute of Paris was organised and operated. Such information formed the body of the submission, which proposed that a scientist of the calibre of the Pasteur Institute's Chamberland should be placed in charge. With a sense of urgency, it was suggested an immediate start could be made by taking advantage of Loir's presence in the colony.

Despite the former conference unanimity, the ensuing intercolonial correspondence soon revealed that not all colonies were equally enthusiastic.[8] The Premier of Victoria felt that until all the financial details were clarified any discussion was premature. The Prime Minister of South Australia considered the formation of such an institute could be better dealt with by a Federal Parliament. Consequently, it was only the Chief Inspectors of Stock of New South Wales, Queensland and Tasmania who met on 15 June 1891 to discuss the establishment of an Australasian Stock Institute, along with a uniform code of regulations for preventing the introduction of foreign stock diseases. By 5 February 1892, Bruce was recommending that the institute go ahead without the support of what he rationalised as the less significant stock colonies of Victoria, South Australia and Western Australia. Suggesting a salary offer of £800 a year to Loir, he proposed that 'a commencement be made at Rodd Island immediately on his return from Europe'.[9]

In general, there was considerable support for the move. *The Pastoralist* bemoaned the delay in establishing an institute based on the discoveries and methods of Pasteur, while the *Australasian Pastoralists' Review* foresaw that the institute 'should become an educational factor and assistant' and was expansive on the benefits to be gained from the changes in 'the methods by which human knowledge have been advanced'.[10] In particular, the urgency of placing the growing meat export industry on a sound basis, with a reputation for quality, warranted the presence of a stock institute which would guarantee disease-free herds and meats fit for human consumption. Even outside the pastoral sector the benefits were obvious, the *Sydney Morning Herald* predicting that the institute could become 'one of the most valuable institutions in the colonies', since 'Australia seems designed to be a land of pastoral production'.[11]

Among those whose interests were threatened, enthusiasm was more restrained. The considerable public pressure in surrounding localities to have Rodd Island returned to public recreation brought on numerous questions in Parliament which raised concerns for safety and suspicions about Pasteur's representatives. Feeding the flames once again were those who supported the idea of a research institute but wanted it to be a wholly Australian institution, having nothing to do with the Pasteur system.

At the forefront, Dr William Camac Wilkinson pleaded for local control and scientific independence, asking 'How long are we to depend entirely on the results of work carried on in Europe? Surely we have here abundant opportunities to contribute to those scientific investigations, which must be of enormous practical value to our special industry.'[12] In a similar mood, the Board of Health also showed itself quite uncooperative toward the implementation of the proposal. Dr Andrew Ross MP, medical practitioner, and botanist, was a trenchant critic of animal vaccination, as we have seen; he never lost an opportunity to raise questions in the House about the activities of Pasteur's representatives in Australia, and called for assessment of the whole proposal by a board of 'thoroughly practical scientific men'.[13]

Meanwhile, Queensland pastoralists were becoming impatient for the benefits of the work of Pasteur trainees. They agreed that Australia should not remain dependent on other countries for research on its stock diseases, but neither should Queensland be dependent on the 'whims and mercy' of other colonies.[14] Proposing that Queensland go it alone, in August 1892 they formed the Queensland Stockbreeders' and Graziers' Association, one of its objects being to secure a stock institute for its own colony, with Loir at its head.[15] Though the Chief Secretary warned that he must respect any intercolonial obligations, by early 1893 it was clear there were none. The intercolonial project had foundered on colonial parochialism and the opposition of vested local interests in the host colony. Queensland was free to go ahead with its own plans.[16]

The services of Loir were no longer available, however. Reportedly 'disgusted' at the lack of appreciation of his gratuitous efforts, Loir had recently returned to Paris and an offer of £500 per year failed to entice him back.[17] The position went to C. J. Pound, 'an expert from King's College, London', who had been in the temporary employ of the New South Wales Board of Health during 1893. Though Pound certainly conducted research over the following years, there seems to have been no associated teaching and there was always a struggle for funds. By the end of the decade the compromise of the Queensland Institute had virtually faded away.[18]

Back in New South Wales another compromise eventuated, after long delays. It too resulted in the substantial separation of research and teaching. When the Stock Institute failed to materialise, demands became more narrowly focused. The industry would settle for a government bacteriologist who could develop and produce animal vaccines and, in particular, secure the theory and practice of anthrax vaccination permanently within the public sector where it could be supplied at a nominal price. With waning interest in the Pasteur vaccine, the campaign led by the stock boards' recently appointed Council of Advice swung its attention to government purchase of the McGarvie Smith and Gunn manufacture. By early 1898 some formal proposals and offers were being mooted.[19] McGarvie Smith and Gunn were not pleased with these predatory grasps at their livelihood, but it was their work which was now seen as vital to the welfare of the pastoral industry.

Two years later, the chairman of the Council of Advice had to admit defeat on this specific issue,[20] but one advance had been made. For bacteriological work the Stock Branch would still have to rely on the services of the bacteriologist attached to the Board of Health, but in 1898 an additional veterinary surgeon, dedicated to the work of the Stock Branch, was employed. This was James Douglas Stewart, an Australian veterinary surgeon trained at the Royal (Dick's) Veterinary College in Edinburgh. There he had won, among other awards, the Silver Medal in Pathology and Bacteriology. Stewart went on to become central co-ordinator of the New South Wales disease control system, as Acting Chief Inspector of Stock in 1904, and Chief Inspector in 1907.[21]

In the meantime, Stewart called regular attention to the urgent necessity for expert and prolonged research on numerous local stock diseases. This, he said, was currently hindered by the lack of a well-equipped laboratory and a properly qualified chemist, animal pathologist and bacteriologist. Stock boards lobbied on similar lines.[22] The New South Wales government finally hit upon the solution of a Government Bureau of Microbiology, established in 1908 under the direction of Dr Frank Tidswell. Here the microbiological expertise of the government services was pooled in a program which combined routine examinations and applied research serving a wide range of industries.[23] For the investigation of stock diseases, facilities were erected on Milson Island in the Hawkesbury River, just north of Sydney.

As for the teaching of veterinary bacteriology, as mooted in the proposals for the Stock Institute, an inadequate first attempt at veterinary education by Sydney Technical College was little improved on by its provision in 1905 of another elementary course.[24] It was the bequest of a benefactor, David Berry, which provided the funds for a Veterinary School at the University of Sydney, opened in 1910. Here, James Douglas Stewart became the first Professor of Veterinary Science, appointed in May 1909, and one of the first intake of students was his nephew, (John) Roy Stewart.

Insisting on a curriculum including 'extensive training in pathology and bacteriology',[25] Professor Stewart saw the appointment in 1911 of Sydney Dodd, D.V.Sc., as lecturer in veterinary bacteriology and pathology. It was only now that research combined with teaching, as Dodd embarked on a program of investigation of animal disease.[26] Stewart himself gave lectures on anthrax, telling his students of the classical method of Pasteur, first demonstrated in Australia in 1888. He went on, however, to indicate that since then McGarvie Smith and Gunn had developed a single-dose vaccine which 'ousted that of Pasteur's'.[27]

An emerging Australian sub-system

The two local investigators who achieved those all-important improvements to the Pasteur vaccine traversed a long learning curve on the way. From simple monitoring of Pasteur's vaccine, a deepening process of engagement with bacteriology and its application led to more and more complex adaptations of what foreigners had to offer.

John Gunn: from pastoralist to bacteriologist

Gunn's first serious bacteriological work can be traced back as early as 1890, when he started monitoring the effects of the first season of Pasteur vaccination. To improve on the rudimentary knowledge he had acquired at the Junee demonstration in 1888, he devoured all the works on bacteriology he could obtain. He also honed his microscopy skills on his imported German microscope, a C. Zeiss Jena. 12213.[28] It was top of its class in its day, and he is said to have became an expert in its use as he graduated from plants and insects to micro-organisms.

Gunn's financial resources were limited, but his improving spirit was given free rein as he built his own laboratory at South Yalgogrin.[29] Here he installed a large vat with a huge brick chimney. In the vat where he prepared broth for growing cultures, he boiled down old sheep from the property and filtered the 'soup' through blotting paper. In devising an incubator for his cultures, Gunn converted a chicken brooder. Though it had a water-jacket, it had no thermostat control and was heated by a small lamp, hand regulated as necessary.[30] Outside, in special pens, he kept some old ewes as experimental flocks on which he tested the Pasteur vaccine for use on Goldsbrough's various properties.

Though Gunn later claimed that he had to work out a method of vaccine preparation unaided,[31] it is certain he would have made the most of any opportunities to learn from Pasteur's representatives. During frequent visits to discuss anthrax vaccination with Devlin, one such opportunity arose from Loir's need for some field trials, probably in 1891. Devlin had urged Loir to explore the possibility that a different strain of bacilli might overcome the mortality problems, and it was arranged that these tests should be carried out by Gunn since Devlin had no suitable place.[32]

A later claim that Gunn spent an apprenticeship in Loir's laboratory at Rodd Island[33] was specifically denied by an employee and long-time friend.[34] From a letter of January 1894 it is clear, though, that Gunn had made an arrangement to carry out the preparation of anthrax cultures for the Pasteur Institute after its eviction from Rodd Island at the end of 1893. As part of the agreement he was to be supplied with 'the necessary germs' and 'every information as to the technical work of the Laboratory'.[35] There is no evidence to suggest that this arrangement went ahead, but there were other avenues to French knowledge.

The most direct was his relationship with Devlin. It is not difficult to imagine Gunn having conversations with Loir at Devlin's home. Perhaps Loir travelled to the Yalgogrin property for the field trials. The generous Devlin also lent Gunn his own vaccination records as well as some original notes on anthrax vaccine sent as a mark of appreciation from Pasteur. No doubt Gunn also obtained some of Pasteur's published work. There is an apocryphal story handed down in the Gunn family describing how he locked himself away for four months while he learned to read French. He was no doubt helped by his wife and laboratory assistant, Jessie, who was fluent in five languages.[36]

Perhaps Gunn was not so diligent about legislation, for the hasty passage of the *Animals Infectious Diseases Act 1888* (described in chapter 4) had passed him by. When he approached the Minister about his work, he was informed that he had no right to carry on as a bacteriologist as he had no qualification and no licence. The Board of Health was horrified to learn that a mere station manager was handling such a dangerous fluid, but after much correspondence was finally persuaded of Gunn's competence.[37] On 27 August 1891, the Department of Lands issued a licence to introduce, keep and inoculate with 'Pasteur's Vaccine de Paris'.[38] In 1892 the licence was extended to investigations on infectious diseases in rabbits.[39]

In soaking up information from all available sources, Gunn was obviously trying to come to grips with the work of major bacteriologists in Europe. His various letters to newspapers demonstrate by late 1891 an ease and confidence in discussing scientific aspects of anthrax vaccination. In February 1892 he made it clear in his correspondence to the *Australasian* that he was familiar with Koch's work on the preparation of pure cultures, particularly of anthrax.[40] He also referred to correspondence with Klein, a major bacteriological authority in Britain, and reported testing some of his suggestions, which he found untenable.[41]

Clearly a determined and dedicated student, Gunn worked 'practically day and night' for two years, devoting all his spare time to his work on anthrax.[42] As his original vaccine was modelled on Pasteur's, the great challenge was to attain the correct *relative* strengths of the two doses, and by October 1892 he had succeeded in producing two standardised attenuations of the anthrax bacillus.[43] The year 1893 brought small-scale tests of his preparation, followed by a trial on 200 weaners. This produced such uniformly good results that, with permission from

Goldsbrough Mort, he went ahead in late October 1893 with the vaccin-ation of 12,100 hoggets, lambs and rams, confident that his vaccine would 'prove equal or superior to the results obtained from Pasteur's'.[44]

Not a single death occurred, either from the effects of the vaccination or from the subsequent challenge with fully virulent bacteria. As he vac-cinated a total of 50,000 sheep in the summer of 1893-94, Gunn also carried out his own version of the Pasteur Junee demonstration, at several stations testing the efficacy of his vaccine by inoculating both vaccinated and unvaccinated sheep with a virulent culture. At Warri Station at Coo-lamon, 3770 sheep were vaccinated at a cost of £23 12s 3d and the manager was delighted at the results.[45] There and elsewhere, as at Junee, the unvaccinated sheep succumbed while the vaccinated sheep were unaffected. In 1894 Gunn submitted his results to Professor Anderson Stuart, president of the Board of Health and of the Royal Society of New South Wales, who judged that 'from the specimens of his work submitted to me [he] appears to be an excellent worker'.[46]

Buoyed up by his results, and planning from early 1894 to go ahead with commercial preparation of the vaccine, Gunn now devoted his attention to finding a technique to preserve the attenuated virus for at least ten to fourteen days. By June 1894 he believed he was 'on the right track'.[47] Along with this went preparations for establishing the frame-work of a competing technological system. First came finance. The pros-pects of financial and organisational support from Goldsbrough Mort had evaporated with their recent forced liquidation and reconstruction,[48] a result of the 1890s depression. So Gunn moved on to another useful connection.

His close friend and neighbour was the nephew of Hastings Cunning-ham, a major wool-broker and financier, who had also been a former mortgagee of Yalgogrin, first as an individual and then through the company, Australasian Mortgage and Agency Co. Ltd, which he floated in Britain in 1880.[49] Its head office was in Edinburgh, but two-thirds of the shares were reserved for Australia. On visits to his nephew in Nar-randera, Cunningham showed some interest in Gunn's proposed scheme. The idea was to float a company called Gunn's Anthrax Vaccine[50] and a relationship with the Australasian Mortgage and Agency Co. offered the possibility of tying Gunn's vaccine business into one of the major financ-ing networks of the Australian pastoral industry. But it too was shaky in the aftermath of depression, and by early 1895 these negotiations also fell through.

On the operational side of the business, Gunn was having more success, progressively building that most vital asset, a competent team of vaccinators. A total of seven assistants, including an agent for Victoria, were trained and licensed to carry out inoculations for the 1894-95 season.[51] Back home in the lab, his 'bright-minded' and devoted wife tended all the incubating cultures. Even so, the sudden explosion of demand took Gunn somewhat by surprise and put considerable strain on equipment and resources. The greater distances to be travelled also put a strain on the stability of the vaccine, causing more failures than for the previous season.[52]

It was at this time, as we have seen, that McGarvie Smith appeared on the scene with a method for 'preserving the special properties of germ products generally'.[53] The partnership with Gunn took effect on 5 April 1895. By the end of April the preservation process had not been perfected but a series of trials was in progress. The priority of the partnership was to overcome the problem of distance and environment *either* by opening branch laboratories in affected districts, or by perfecting preservation.[54] As it turned out, the technical solution rendered unnecessary an organisational response to geography and climate.

John McGarvie Smith: from watchmaker to bacteriologist

John McGarvie Smith was a city man, born at Paddington, Sydney, in 1844.[55] Obliged at the age of thirteen to seek his own livelihood, he learned the trade of watchmaker and jeweller, but he was a man of wide interests, both sporting and intellectual. In his youth, McGarvie Smith captained the first Australian rifle team to compete in America. But it was an attraction to photography that led him to chemistry, and at twenty-three he began attending the classes of Professor John Smith at the University of Sydney in his spare time. These classes were open to non-matriculated students and there McGarvie Smith drew scholarly succour for seven years. In 1874 he joined the Royal Society of New South Wales. He had by then progressed into geological and metallurgical studies under the new professor, Archibald Liversidge. He later claimed to have devised treatments for refractory ores, reportedly solving problems on several mining fields including Sunny Corner, Broken Hill and Mount Morgan.

McGarvie Smith seems to have maintained his jeweller's shop in the city, but whether by watchmaking or metallurgy, he is said to have amassed considerable wealth. He and his wife Adelaide, the widow of the intellectual and political Daniel Deniehy, lived in some comfort in the fashionable suburb of Woollahra, more prestigious than the adjacent Paddington. By the time he was forty (1884) he had the means to tour Europe where he furthered his scientific training, studying 'his favourite subjects' in France and Germany and extending his chemical knowledge by visits to German works with his friend Dr John Elliott of Elliott Bros, the Sydney drug wholesalers.[56]

It was Elliott who later encouraged McGarvie Smith to take up bacteriology. This time he could not resort to the university, for bacteriology was not yet taught there as an individual subject. It was fortunate, then, that Oscar Katz, Ph.D., was in the colony, working under the patronage of William Macleay, and also as expert to Melbourne's Sanitary Commission and then Sydney's Rabbit Commission. In late 1889, McGarvie Smith engaged Katz to teach him bacteriology for six months; he said he paid a very high salary.[57]

Such was Smith's introduction to the Koch traditions. Soon Smith was engaged in serious bacteriological experiment. Soon he too came under the nervous scrutiny of the Board of Health, who learned at the end of 1890 that he was working with the bacilli of typhoid fever. But after an interview with the president, and an inspection of his home laboratory at Denison Street, Woollahra, Smith received the licence which sanctioned his use of a variety of microbes, including anthrax. Within a few months he was in trouble again, this time for injecting the microbes into experimental animals. The board said his home premises were not sufficiently isolated for work with live animals—he was a danger to the public health. But Smith had connections and the Minister on his side, and by mid-1891 he had a licence to inoculate rabbits with the microorganisms of typhoid, anthrax and tuberculosis.[58]

McGarvie Smith was fortunate to have the financial resources to support his study. His laboratory was described in 1892 as 'a suite of such establishments, all most perfectly equipped with the most modern of appliances'. It also had a library of scientific works considered 'priceless in value'.[59] He had already displayed his bacteriological techniques and cultures to the Royal Society of New South Wales in 1891,[60] and by the time of the Intercolonial Medical Congress of Australasia in 1892, McGarvie Smith was sufficiently confident and competent to contribute to the collection of exhibits organised by Dr William Camac Wilkinson in his laboratory at the university. Here McGarvie Smith provided an 'excellent display of pure cultures of the tubercle bacillus'.[61] By late 1892 he must have achieved some professional recognition, for in November he was commissioned by the Metropolitan Board of Water Supply and Sewerage to undertake a bacteriological investigation of the air in its sewers.[62] In this he was assisted by C. J. Pound, who was destined to become the future Government Bacteriologist in Queensland. The subsequent intense public debate about the results clearly did no harm to McGarvie Smith's reputation, for in 1894 he carried out sanitation tests on Sydney's wood paving, overturning the report of an earlier Royal Commission by finding no trace of germs injurious to life.[63]

The wide-ranging McGarvie Smith had another useful interest, probably also passed on by Oscar Katz. For Katz had taken up the subject of

Australian snake venom in 1888 at the urging of William Macleay.[64] Now McGarvie Smith was intrigued by the constitution of snake venom and the possibility of developing a vaccine. So too was Charles Martin of the University of Sydney, and the results of their collaboration in the university's physiological laboratory and McGarvie Smith's home laboratory were presented in a joint paper to the Royal Society of New South Wales on 3 August 1892.

It was this work which probably brought McGarvie Smith into the realm of the Pasteur tradition, for Loir was certainly interested in the effects of snake venom and the preparation of anti-toxins, as was his colleague, and then successor, Dr Louis Momont.[65] In 1894, with Rodd Island no longer available for the work of the Pasteur Institute of Australia, McGarvie Smith offered Momont the use of his laboratory free of charge in return for bacteriological instruction which would add the Pasteur tradition to his expertise.[66] Perhaps it also put him on the track which led to partnership with John Gunn.

The partnership

Just when and how McGarvie Smith and Gunn came together is the subject of conflicting stories. Gunn's neighbours said he had succeeded by 1894 in producing a single-dose vaccine but was unable to repeat it and enlisted McGarvie Smith's help, via Arthur Devlin.[67] From the Pasteur camp came the story that Gunn could not fill vaccine tubes aseptically and was brought in contact with McGarvie Smith when another Goldsbrough Mort manager discovered Smith had the required techniques.[68] Another story told that it was simply Gunn's need for vaccine containers which led him to 'an authority' at the university, and on to McGarvie Smith.[69] Yet another version held that the two men sought each other out since 'Gunn had all the practical knowledge and considerable theoretical skill, McGarvie had little practical, but a thorough insight into Pasteur's methods'.[70] McGarvie Smith himself insisted he had instructed an ignorant Gunn 'in the preservation of spores'.[71]

Whatever the true story, it is easy to imagine ways in which the self-taught pastoralist and the city scientist could have crossed paths in the small world of 1890s science. Perhaps the explanation lies with Gunn's correspondence with Professor Anderson Stuart at Sydney University at a time when McGarvie Smith was collaborating with Charles Martin, a member of the same medical faculty. Stuart was also president of the Royal Society of New South Wales, of which McGarvie Smith was a member, and Stuart's anniversary address to the society in May 1894 mentioned both Gunn's work on anthrax and Martin's work on snake venom.

Perhaps there was a more direct link straight through the Pasteur Institute of Australia. Gunn's field trials for Loir were followed by further trials for Momont who was working in McGarvie Smith's own laboratory. Devlin, the Pasteur agent, was also a vital line of communication. Leslie Devlin recalls how his father's proposal for a single-dose vaccine had been rejected by Loir and then Momont, but seized upon by an eavesdropping McGarvie Smith who prepared some experimental material. Its failure in tests conducted by Devlin, probably in 1894, led McGarvie Smith to suggest a joint venture to produce and market a vaccine, an offer which Devlin declined; presumably John Gunn did not.[72]

It seems fairly certain that what McGarvie Smith had to offer Gunn was finance and a way of using the antiseptic properties of glycerine to protect his vaccine. McGarvie Smith later also claimed that it was he who 'possessed the knowledge (and demonstrated it) in giving immunity by one inoculation', even prior to Momont's sojourn in his residence.[73] Professor Anderson Stuart believed it was Gunn who was responsible and that the irascible McGarvie Smith 'hated him ever afterwards'.[74] Government Analyst, William Hamlet had another theory: having witnessed 'the great hatred' between Oscar Katz and Pasteur's delegates, he speculated that Katz had stolen Loir's detailed description of the method of attenuation and sold it to McGarvie Smith.[75] What seems much more likely is that Katz passed on to Smith the Koch techniques of preparing cultures on solid media, and the preservation of spores on silk threads. These would form part of the McGarvie Smith and Gunn methods, but were not practised by Pasteur or his delegates.

Whatever the original source of the single-dose method, it is evident that there was some combined experiment on Gunn's vaccine before the partnership was mooted. An interested neighbour who visited Yalgogrin often to watch the progress of these experiments later recalled how Gunn expected McGarvie Smith to retire from the scene, but, now knowing the details of Gunn's methods, Smith made it clear he was determined to share in any commercialisation. Gunn was wary, having heard how McGarvie Smith tricked snake sellers into accepting a low price by getting them to release the snakes onto a slippery polished floor and then rejecting them.[76] McGarvie Smith certainly seems to have been wily, as well as eccentric, laying foundations of mistrust for the partnership. Perhaps this is why the deeds specified that each man work entirely independently, producing his own vaccine in the most convenient manner, but communicating all knowledge to the other partner so both would be on the same footing.[77]

Once the partnership had been agreed upon, the two men entered into a series of tests to verify the efficacy of the emulsion which resulted

Figure 6.1   McGarvie Smith milking his snakes for venom (*Sydney Mail*, 24 December 1892)

Figure 6.2   John Alexander Gunn, MLC (Jack Gunn, descendant)

from McGarvie Smith's idea of mixing glycerine into the diluted broth culture of the anthrax bacillus which Gunn had formerly injected into the animals. In the first of these tests, twenty sheep were injected with vaccine bottled for five weeks at a temperature of 90°F. Under the old method the vaccine would have been inoperative after the fifth day, yet the treated sheep all withstood the challenge test with virulent anthrax without effect, while an unvaccinated sheep died within 24 hours.[78]

Parallel research was focused on the search for the method of producing a single-dose vaccine. The firm's vaccination returns show that several thousand sheep on Yalgogrin were given only one inoculation in the 1895–96 season. Experiments continued over two years and 25,000 sheep.[79] By mid-1897 they were sufficiently confident to advertise the availability of the single-dose vaccine in their brochure, and in the 1897–98 season it was put into general operation. The records submitted to the government from this time did not include mortalities, making it difficult to specify the success rate of this season, but Gunn claimed that over one million sheep were 'protected by a single inoculation'.[80]

The measure of scientific success

All acknowledged the commercial success of the McGarvie Smith and Gunn vaccine, but what did it represent in terms of scientific achievement? Assessment is hindered by the secrecy of their work and the lack of any published details, as well as the confusion of conflicting claims. Dr Frank Wall, one of the two people to whom McGarvie Smith confided his formula just before his death, was convinced that it 'upset all modern notions of bacteriology on the subject'.[81] Roy Stewart, nephew of the Professor of Veterinary Science at Sydney University, and a distinguished student of veterinary bacteriology, claimed after four months' training in McGarvie Smith's laboratory in 1915 that he did not know of any other place where he could obtain such knowledge.[82] On the other hand, Professor Anderson Stuart, as a confidant of Gunn, believed their process to be 'a very simple matter', involving no profound knowledge of the science of bacteriology.[83] Nevertheless, he thought it probably applicable to the production of other vaccines for animals and humans. The details, he asserted, should therefore be published through the proper scientific channels.

They were not, but a detailed analysis[84] of what is known of their methods and of the contemporary knowledge and techniques of bacteriology reveals that, though no advance in theoretical knowledge had been made, the McGarvie Smith and Gunn method was more than a copy of an existing process. It had drawn on the Pasteur and Koch traditions and techniques, blending them in a unique way to develop a product with quite specific qualities.

With regard to attenuation, the challenge for the Australians was to find a level of virulence which would render a single dose safe, conferring in one shot the immunity previously reached more gradually in two doses. For this it had to be just sublethal for sheep, and therefore somewhere between the strength of Pasteur's first and second dose. McGarvie Smith said he had achieved the appropriate level of attenuation by cultivating fully virulent bacilli in the abdomen of a fowl.[85] However, it was a combination of two different aspects of Pasteur's work which provided the basis for the firm's method as described by Gunn. The logic was to weaken the bacteria, then strengthen them.

The first step was to hold anthrax bacilli in meat broth at 42°C until the preparation was found not to kill sheep, 'probably around the 24th day'; it was then approximately equivalent to Pasteur's first vaccine dose, and too weak for the single-dose vaccine. Pasteur had shown however, that weakened bacteria could be returned to their original virulence if they were made to pass through a succession of various animals. Gunn found that the appropriate strength could be reached by passing the weakened anthrax culture through mice, then young guinea pigs, then older guinea pigs, then through young rabbits, and finally older ones. For Gunn, it was 'a long and intricate process'.[86] But the achievement lay in applying Pasteur's principle to target a particular level of attenuation, and finding the combination of animals which would produce it.

The second element of the McGarvie Smith and Gunn method was fundamental to the first. It lay in the development of production techniques which would maintain the potency of the vaccine at this critical level even under the harshest environmental conditions. From Koch and Pasteur, they knew that spores were the key to vaccine survival. In 1876 Koch had discovered the endospore phase and its role in preserving the virulence of the anthrax bacillus; by 1881 Pasteur had reported that attenuated microbes could also form spores which would fix their strength at that level;[87] later that year he tested these qualities in transporting the vaccine over long distances.[88] McGarvie Smith and Gunn drew on this knowledge and cultivated attenuated bacilli until the spore stage was reached, so that the vaccine, as distributed for inoculation, contained virtually all spores rather than the more susceptible rods. The highly resistant spores were unaffected by the extreme climatic conditions experienced in inland Australia and were crucial to the safety and viability of a single-dose procedure. The use of spores was not original, but in applying these spore properties to the preparation of a single-dose vaccine, the Australians led the world by many years. Though others would eventually follow a similar path, and spore vaccines became the most widely accepted form of vaccination in the twentieth century, the first reports of a single-dose spore vaccine did not appear until after 1910, more than a decade after McGarvie Smith and Gunn's release.[89]

Pushed by Australian conditions, guided by first-hand exposure to European bacteriology through the scientific networks of Sydney, McGarvie Smith and Gunn were able to screen existing strands of bacteriological knowledge and bring them together to adapt the Pasteur vaccine. In the process they created a product which was unique for its time. But they did not stop there. Growing confidence as scientific experimenters spurred them on to a search for the agent responsible for providing the immune reaction. In June 1897 McGarvie Smith reported success in isolating this vital agent[90] and in August 1899 there appeared in the *Australasian Medical Gazette* an article detailing experiments by McGarvie Smith and Gunn which aimed to develop a cure for animals already infected with anthrax.[91] Gunn had since been moved to Borambola station, on the Murrumbidgee, and here in a new laboratory the serum of sheep immunised by prior vaccination was injected into infected animals. Three series of experiments aimed to discover the relationship between the subsequent course of the disease and three particular parameters—the size of serum dose, the number of doses, and the time lag between anthrax infection and serum inoculation. The results encouraged the conclusion that the curative properties of the serum were greater than had 'heretofore been accredited to it by other observers'. It was hoped that ultimately the serum might be used to cure anthrax in humans, and that the firm would be able to produce the material in its laboratories.

Investigations of immune serum of various kinds were under way in several parts of the scientific world,[92] and these experiments indicate that Gunn and McGarvie Smith were now screening the scientific literature more extensively and attempting to branch out in their research activities. There is no evidence of this venture coming to any practical conclusion, but further modifications to the anthrax vaccine process were made. In 1912, two years after Gunn's death, McGarvie Smith had what he described as an 'inspiration'. The result was what he called the *matured* spore vaccine.

In essence there were two modifications derived from the 'inspiration'. The first reduced the period of incubation of the attenuated culture and resulted in the complete degeneration of rods to spores, the 'matured spores' as they were called by McGarvie Smith. The second modification produced a glycerine-water emulsion of standardised opacity, the degree of opacity being the measure of the concentration of the spores and a simple means of control.[93] This would certainly, as McGarvie Smith pointed out to the Chief Inspector of Stock,[94] remove the great toil of testing a range of concentrations on sheep before choosing the most appropriate. As such, the modification represented a significant advance in process control. Again other scientists in the world were working on similar lines[95] but again the Australian developments seemed level with, or ahead of, those elsewhere. All the evidence is suggestive of innovation in the application of established knowledge and techniques.

Establishing institutional roots

By 1898 the locally modified McGarvie Smith and Gunn vaccine, specifically adapted to the idiosyncrasies of Australian conditions, had virtually displaced the Pasteur vaccine from the Australian market. A competitor which attempted in 1906 to 'dispel the monopoly' was soon dispelled itself.[96] As a profitable business built up, so did the signs that a competing local technological system was developing as a separate offshoot from the Pasteur system and would develop some technological momentum in its own right.

One of the signs was the publication, already mentioned, of the report of McGarvie Smith and Gunn's anthrax research in the *Australasian Medical Gazette*, a serious medical journal, in 1899. In the same year came a request from the Victorian Chief Inspector of Stock for samples of the vaccine to be supplied to Dr Cherry, the bacteriologist at the University of Melbourne, for comparative tests with the Pasteur Institute product.[97] Soon after, the Victorian Chief Inspector gave official recognition to McGarvie Smith and Gunn vaccination certificates as sufficient authority to permit stock to cross the Victorian border.[98] Victorian inspectors of stock even began the practice of arranging for small stockowners to combine to take advantage of the firm's discounts for large lots.[99] Similar accreditation of McGarvie Smith and Gunn vaccination was also recognised for import of Australian stock into New Caledonia.[100]

Meanwhile Gunn's authority grew as a person of some bacteriological acumen. When the new century brought a renewed approach to Paris for a biological method of rabbit control, John Gunn was appointed chairman of the Rabbit Destruction Fund Committee which arranged and supervised the whole endeavour. By 1905 he was a successful grazier with his own property, Braehour, but also described himself as a bacteriologist.[101] Gunn's stature also grew generally within the community. By the time he died in 1910, he had been chairman of the Wagga Wagga Pastures Protection Board, chairman of the Stock Board's Council of Advice, a member of the Kyeamba Shire Council, vice-president of the Murrumbidgee Pastoral and Agricultural Association and a member of the Council of the Stockowners' Association of New South Wales, all of which gave him enormous scope for building the credibility of and commitment to the McGarvie Smith and Gunn system. In 1908 he was appointed to the Legislative Council.[102]

Despite all this, the secrecy and the enmity which developed between McGarvie Smith and Gunn constrained the development of secure institutional roots for their methods. Their production techniques had been neither patented nor reported in technical journals; they were sealed in the memories of two mortal men. The partnership deed specified that neither could sell his share to an outsider,[103] preventing any other commercial access to the information. And all the time, the antagonism between the two men was getting worse: by 1906 Gunn described McGarvie Smith as 'a devil neither more nor less and a mad one at that but with the cunning of the devil himself'.[104] By 1907 the partnership was in the hands of an arbitrator.

As the relationship with McGarvie Smith deteriorated, Gunn said he feared for his life, and he sealed the description of his method in an envelope, to be opened only after his death to allow his family to carry on the business.[105] It was offered instead to the New South Wales government by the trustees of his estate. It was, however, regarded as unnecessary by a Chief Inspector of Stock who had come to office after anthrax was under control: he said the Pasteur Institute in Paris distributed vaccines all over the world, but anyway he favoured a proposal recently approved by the Minister for Agriculture for the government to establish its own experimental station under a veterinary pathologist, whom he assumed would be able to prepare an adequate vaccine.[106]

The stockowners were of a different view, adamant that it was 'absolutely essential for the preservation of the flocks of the State to retain the secret preparation of the vaccine'; none other gave such satisfactory results.[107] They were also nervous about the fact that the secret now resided with only one man, who was old, obviously in poor health, and unable to keep up the vaccine supply. They wanted the government to 'appoint a scientific man to be taught the art of preparing the vaccine by Mr Smith'.[108]

As was his cantankerous way, McGarvie Smith chose instead to instruct Roy Stewart, who now had a degree in veterinary science and lectured on animal husbandry at Sydney Technical College. He came from a long line of veterinary surgeons, being the son of John M. Stewart, veterinary surgeon and proprietor of the practice of John Stewart and Sons, and grandson of the late John Stewart, a graduate of Edinburgh. His Uncle James, as we have seen, held the chair of veterinary science at the University of Sydney. At this time Roy Stewart was employed as a bacteriologist in John Stewart and Sons which, under the name of Biological Stock Institute of Australasia, had been supplying a number of vaccines for a few years, including a bouillon anthrax vaccine.[109]

The instruction from McGarvie Smith began in November 1915 and on the basis of what was passed on, the Stewart family built a modern new laboratory at Randwick, at the cost of more than £1000, to manufacture the matured spore vaccine. Here yards were constructed for the testing of each brew and a fly-proof room where inoculated sheep were injected with a challenge anthrax culture. McGarvie Smith collaborated on the design of the laboratory and supervised the manufacture, by John Turnbull of Argyle Cut in Sydney, of two sterilisers to Smith's own design. He also sold the Stewarts one of his own anthrax incubators and supervised its installation.[110]

Yet this passage of expertise was neither simple nor untroubled. McGarvie Smith had at one time suggested that he would hand over to the Stewarts the goodwill of his business and the name of McGarvie Smith Matured Anthrax Spore Vaccine to go with the method of preparation.[111] But by 1916 it was clear that McGarvie Smith had antici-pated work recently published by Eichhorn in the United States, and the Minister for Lands, W. G. Ashford, was impressed. He offered to establish a McGarvie Smith Institute under departmental auspices. There was also talk of a knighthood.[112]

Here was the chance for immortality and glory. After a cat-and-mouse game with the government which lasted right up to 1918, McGarvie Smith finally handed over the method of preparation on his deathbed. It went into the hands of two government agents, on condition that it remain a secret known only to an institution to be named the McGarvie Smith Insti-tute. This foundation, to which he bequeathed £10,000, was to be con-trolled by a board consisting of representatives of the pastoral industry and representatives of the government. Only these trustees would have access to and control over the process, and not the government itself. The profits obtained from vaccine sales were to be used to defray the expenses of the institute, to support research on animal diseases and vaccines and medi-cines, and to assist veterinary science in general.[113]

Kept in a bank vault in Sydney, the method and its companion sample remained the secret of the trustees and bacteriologist of the institute, which fulfilled the conditions of its charter in a number of ways. Schol-arships assisted students graduating in veterinary science. Donations con-tributed to the provision of research facilities at the University of Sydney, the Glenfield Veterinary Research Station and the New England Nutri-tional Research Station. Other funds supported travel to overseas confer-ences, and post-graduate research. In 1927 the Board of Trustees of the institute appointed the first Research Scholar, Grahame Edgar, who carried out research on behalf of the institute at Glenfield, and became the Director-General of Agriculture for New South Wales. By the 1960s the McGarvie Smith Institute had contributed nearly £100,000 to various veterinary research programs.[114] By 1992 dispersed grants totalled $540,365, worth over $5.6 million on current values.[115] The funds had come from vaccine sales within Australia and also overseas, especially to the lucrative South American market.

Gunn's method took a different institutional route. It was stated in 1915 that John Gunn's son Angas was intending to take on the anthrax business when he came of age in 1916,[116] but eventually Gunn's widow was galvanised into action by the many statements about McGarvie Smith's intentions. Anxious to preserve her husband's name in the records of discovery, on 10 August 1918 she again offered to hand over Gunn's formula free of charge to the government, provided due recog-nition was accorded by perpetuating his memory.[117] This obligation was discharged by giving Gunn's name to a wing of the new Glenfield Vet-erinary Research Station opened in 1923.

Meanwhile, Roy Stewart manufactured the vaccine, as per the method taught by McGarvie Smith, and marketed it under the label of the Biological Stock Institute of Australasia. By 1918 Francois Ray cut the market into three. Trained at the Pasteur Institute in Paris, he had been resident in Australia from 1906 and for some years had been supplying a number of animal vaccines. He too was now advertising his anthrax vaccine as using 'matured spores only'.[118] In 1928 Ray joined forces with a young pharmaceutical chemist called Arthur Webster and together they produced anthrax and pleuro-pneumonia vaccines, marketed under the Ray and Webster label.

Though these two men parted company after a few years, Webster had learned much about bulk production of animal vaccines. After completing a veterinary science degree during the 1930s, Webster built up a vaccine production business of his own, branching out from stock vaccines to fowl vaccines and becoming one of the few people in the world developing equipment and techniques for the fledgling science of vaccine technology. He became the first scientist to produce fowl pox vaccines using mild and attenuated virus strains; he developed and produced Australia's first commercial vaccines for infectious laryngotracheitis and infectious bronchitis in poultry; and in co-operation with the New South Wales Department of Agriculture and university scientists, he gave Australia its first vaccine for Marek's disease. When the McGarvie Smith Institute withdrew from vaccine production in 1970, it was to the now flourishing Arthur Webster company that the anthrax vaccine manufacture was bequeathed.[119] Meanwhile, in Melbourne the Commonwealth Serum Laboratories, formed in 1916 as a public institution for the development of vaccine research and technology, had become a formidable competitor in the expanding vaccine industry in Australia.

Some reflections

The Australian system of disease control had come a long way during the previous century. At the beginning of the 1880s it still had little in the way of scientific components which were likely to generate internal solutions to disease problems. But, with Alexander Bruce as 'system manager', a process was set in train which developed the ability to search out, assess and assimilate technological solutions from outside. Though the scientific community had hitherto had little involvement with disease control, it now contributed by way of public debate, and provided the networks, institutions and pools of expertise which formed the backdrop to the development of McGarvie Smith and Gunn's bacteriological skill.

This was important, since the imported technology was imperfect. It was the capacity to learn from the French and gradually adapt their technology to local conditions that made the difference between ultimate success and ultimate failure. That learning was passed on to an ever-widening circle of people. At one level, the whole process of generating demand helped raise the scientific consciousness of a significant proportion of the population. At another level, the technology of anthrax vaccination passed into the standard veterinary curriculum, together with the science from which it had developed.

The idiosyncratic and *ad hoc* aspects of the Australian system appeared, however, in the way the local vaccination system developed. Without patent rights and without a structured, formal organisation, the local system remained constrained by the need for secrecy and the lack of financial and institutional backing. The secrecy prevented diffusion of the details of advances made in the process of adaptation. Nevertheless, that process of incremental adaptation had brought full control of the technology from Paris to Australia.

PART III
Rocks, cyanide and gold

CHAPTER SEVEN

Australian gold, British chemists

From feast to 'famine'

It was a mid-century feast of gold which rushed Australia to the fore of the world's producers and helped to push world output to an all-time high in 1853. Technology then was simple—the pan for washing the gold-bearing sand in the water-stream, or the cradle, a long box standing on rockers. Using these, the free metallic gold could be washed and encouraged to settle out from the lighter, surrounding earth.[1]

By the 1870s mining life was more arduous and mining production in decline. The cream of the known alluvial deposits had been skimmed off the surfaces of the world's goldfields, and the auriferous quartz veins running deep into underlying bedrock scorned the simple tools and techniques of sluicing. Progressively more dramatic and expensive methods prised out the gold-bearing rock, as hammers and muscle gave way to drills, dynamite and crushing batteries. But something more than mechanical force was needed.[2]

Amalgamation of gold with mercury had for a time offered some additional extracting power.[3] When finely crushed rock was carried over an amalgamated copper plate by a stream of water, some gold was caught as it formed a heavy, fluid alloy with the mercury. The sandy tailings were washed away, the amalgam scraped off and the mercury recovered by distillation, leaving the metallic gold. If all the gold existed in the free metallic form, recovery would be complete.

But the gold in quartz veins came in two forms. Above the permanent water level natural processes of weathering gradually oxidised and washed away the compounds of sulphur, arsenic and other impurities, leaving the inert metallic gold in the crevices of the equally inert quartz. Deeper down, the atmosphere and weather were powerless, and the so-called refractory ores retained all their minerals untouched and unchanged. It was these ores which were at the heart of the industry's troubles. They bore the collective label 'pyrites', because of the intimate association of the gold with sulphur compounds and other minerals, and the deeper the mine, the more they predominated. At the water level, free gold had often disappeared entirely.[4]

Enclosed in their casing of sulphides of base metals, the gold particles in these ores passed over the copper plates and escaped into the tailings. Concentrating machinery, which called on relative specific gravities to separate the pyrites and other sulphides from the sand, did not solve the problem of how to extract from the concentrates the gold that was not in the free metallic form. Long continuous grinding with mercury in cast-iron pans might yield considerable gold, but lose much of the expensive mercury. The alternative of smelting the concentrates required well-built brick furnaces, coal and fluxes—impracticable for many isolated and uncongenial mine locations.

It seemed that chemistry might come to the rescue when the chlorination process made use of the fact that chlorine gas transforms metallic gold into a water-soluble chloride, whilst it is without effect on most metallic oxides. Yet severe limitations remained. In the competition for chlorine molecules, the relatively small proportion of gold fought the chlorine-hungry sulphides and arsenides of iron, copper, lead, zinc and other metals. Only prior roasting in a furnace could expel the sulphur and oxidise the metals, but the process was still not applicable to lead-bearing ores, and it was always troublesome and never cheap.[5]

The technology for extracting gold from its encasing rock was simply inadequate, and world output showed the deficiency. By the early 1880s the 1850s peak had been cut by over a third.[6] Fear of a possible 'gold famine' was rampant as gold demand continued to soar.[7]

Since Australia was a major gold producer, and one which took its methods from the international stock, Australian goldmining shared this experience. By the 1870s government mining officials were frowning at goldfield inefficiencies and puzzling over the limitations of available techniques.[8] They had the advice of scientific and mining experts but seemed to make no headway. Deep in the earth, modern diamond and rock drills blasted out more and more rock, but at the surface the methods for extracting the gold used inefficient mills and little metallurgy. By the 1880s both production and productivity were plunging to their lowest ebb: New South Wales saw the value of each miner's output fall from £63 in 1881 to £37 by 1890,[9] while Victoria's 614,838 ounces in 1889 was a poor reminder of the 3 million ounces mined in 1856.[10] Nor did the situation seem likely to improve now that quartz mining accounted for 70 per cent of the industry's work.

The problem remained unsolved, despite searches for suitable technology overseas. By 1889, a royal commission set up to investigate the continuing decline in Victoria's production found that 'mining generally became more difficult and expensive every year'.[11] At the same time a special committee of the Australasian Association for the Advancement of Science was conducting a survey of gold-saving appliances.[12]

A British technological system

The Australians did not yet know that in far-away Glasgow, a small research syndicate had joined the race to find a way of loosening nature's grip on her deeper treasures. Toiling through the night in a basement laboratory, chemist John Stewart Macarthur and two physician brothers, Robert and William Forrest, were spurred on by their financial backer, George Morton. As a Glasgow businessman and stockbroker, he knew that 'the world wanted gold, and must have it';[13] perseverance, he said, would be rewarded. He was right. By 1887 they had come up with the cyanide process of gold extraction, which a contemporary ranked 'amongst the most remarkable discoveries of the present century in metallurgical science'.[14]

The starting point of their metallurgical journey was the long-known scientific fact that pure metallic gold dissolves in solutions of alkaline cyanides. The importance of their discovery lay in the revelation that potassium cyanide solutions could leach gold from its many complex physical and chemical combinations, by *selectively* dissolving the gold and leaving behind the associated minerals. This new knowledge was patented as the Macarthur–Forrest process for the use of cyanide to dissolve gold from its ores.

The physical form of this new technical system brought chemicals and ores together in large vats with mechanical agitation. The crushed ore or tailings were treated with a weak solution of potassium cyanide, generally under 0.5 per cent. The cyanide 'substantially' dissolved the gold, while 'substantially' leaving the other constituents of the ore untouched. The gold solution was then drained off, or displaced by water, and made to flow through a mass of zinc shavings or threads, which precipitated the gold as a fine powder while the solution passed on. The precipitation on zinc was the subject of a second patent.[15]

It was true that potassium cyanide was a very poisonous chemical, but so were chlorine and mercury. Many advantages were claimed for this new process, including its suitability for remote locations, since it required none of the furnaces, coal, and fluxes which made smelting expensive and often impossible. In contrast to chlorination, the cyanide process eliminated roasting, and therefore furnaces and fuel. As well, the hitherto unprofitable ores containing lead, zinc, or earthy carbonates could be worked economically. Chlorination used one ton of chemicals to treat only 14 tons of ore, but cyanidation could treat

100 tons of ore with that amount of chemicals, and extract silver as well as gold.

The cyanide process emerged as a radical solution to a problem in gold extraction in general, and the chlorination process in particular. But Macarthur was no builder of systems. He was a small cog in a much bigger wheel.[16] By night he might follow his own dreams but by day he was the chemical servant of the Tharsis Company, which mined the pyrites of Spain, diverting the sulphur, copper and iron back to its several chemical and extraction works in Britain. Tharsis was in turn a subsidiary of Glasgow's St Rollox alkali works and part of the considerable business interests of its master, Charles Tennant, who dominated the British alkali trade.[17] Though discovered through private and independent research, the Macarthur–Forrest process came to be absorbed and developed as a unit of the extensive and established technological system that Tennant controlled.

Charles Tennant was a man of confidence, wide horizons and diverse interests. He headed a network of chemical, mining and metallurgical enterprises which held and worked an impressive range of patents. The links penetrated deeply into the British chemical industry and beyond. The Tharsis directors and large shareholders were an equally notable set of men, collectively responsible for much of the extraordinarily vigorous business initiative of Glasgow. Among them were two frequent co-investors with Tennant, Archibald Shanks Schaw, Glasgow ore merchant and dealer in minerals, and John Wilson, of the Hurlet and Campsie Alum works.

Tennant's move into goldmining arose from the extension of his network to London with his election as a Liberal MP in 1879. Within the year he had formed two goldmining companies in India. Here he experienced first-hand the frustration of knowing that much of the Mysore gold remained beyond his grasp, untapped by a lagging technology. Always on the alert for new and better methods, he was aware that an extraction process that could deal with the pyrites would bring rich rewards from a desperate industry.[18]

The events which brought John Macarthur and his process within the bounds of Tennant's domain began with an American chemist, Henry Renner Cassel, who also had his eye on prospective riches. In 1884 Cassel presented himself in Glasgow, claiming to have found the answer to the gold-extraction riddle by applying electricity to rid the chlorination process of its practical and economic limits.[19] Specifying several technical advantages for his process, Cassel convinced Tennant of its great profit potential.

On 15 December 1884 the Cassel Gold Extracting Company was formed to exploit Cassel's patents, with a capital of £150,000. The subscribers to the Memorandum of Association read like a *Who's Who* of

Tharsis, with Charles Tennant at the head. Cassel was to be general manager, receiving for his patents shares worth £50,000. In the excitement of the vigorous promotion that followed, speculators rushed Cassel shares high on the Glasgow Stock Exchange, and Cassel promised to do for the goldmining industry 'what Bessemer did for iron and steel'.[20]

By the end of 1885 patents covered the process around the world, and in early 1886 expeditions were sent to Australia and North America. Only months later, midway through large-scale testing programs in London, Cassel suddenly left the country after selling all his shares at the top of the market. Behind him remained a process which could not be made to work: in the words of one critic, 'instead of being a cheap method of gold extraction, it was in reality an expensive way of producing bleaching powder'.[21]

It was fortunate, though, that John Macarthur had, in the meantime, persuaded his colleagues to include Cassel's process in the syndicate's research program. Now his article on the subject passed before the eager eyes of a still confident Charles Tennant. At a meeting on 9 November 1886, Tennant and his board agreed that a 'thoroughly practical man who understands and has faith in the process should be engaged to take charge of investigations and to develop the process to perfection'.[22] John Macarthur seemed the perfect choice. As manager of the Glasgow works, he received a salary of £350 per annum plus the dividend on 500 shares.[23] His insistence on maintaining parallel and independent investigations with his syndicate was resolved with an arrangement that the company would have the first option on any discovery which might result.

It was therefore within the framework of this contract that the cyanide process came to fruition. A continuing program of experiments within the Cassel works was designed to retrieve the Cassel process, but this was matched by the quite separate, and extensive, series of syndicate assessments of potential gold solvents. Among these was potassium cyanide. In November 1886, it was tested on some Indian tailings, but did not seem to work. It was only eleven months later that further trials revealed the true potential of cyanide,[24] and on 19 October 1887 patent protection was obtained when a provisional specification of the English patent, No. 14,174, was lodged by John Macarthur, R. W. and W. Forrest.[25]

Within days the Cassel Board had sanctioned application for immediate protection in several other countries, and had completed formalities for their own exploitation of the process. The works were now divided into three sections, one for the Cassel process, and two for the Macarthur-Forrest process. One of the latter was for extraction, the other for recovery and manufacture of the reagent. Though the Cassel section was later

Figure 7.1 John S. Macarthur, Macarthur–Forrest cyanide patentee (*Australian Mining Standard*, 15 February 1900)

discarded, the third proved a significant addition to the Tennant technological system, with a cyanide manufacturing capacity that by the end of the 1890s was reportedly more than double that of any other factory in the world.[26]

During the intensive developmental work which scaled the process from laboratory to plant, the syndicate members scoured all the chemical and technical textbooks. There they could find no information at all on the action of cyanide solutions on gold ore, and no direct or useful information as to the action of cyanide on the ores of base metals. What they did find in the textbooks was some information on the action of cyanide on the metals themselves, but much of this was proved erroneous in practice. Their experiments therefore fell into two sets, one designed to produce a simple, efficient process based on the reaction they had discovered, the other to clarify their theoretical understanding.[27]

On the practical side it remained to devise an industrially applicable form of the fact that zinc precipitated gold from its cyanide solution. After numerous tests on finely divided zinc in various forms, it was found that zinc shavings delivered reliable recoveries, as described above. A patent taken out on 14 July 1888 covered this and several other refinements made since the granting of the provisional protection. One which would become important for the future was the percolation method of leaching, using tanks with permeable false bottoms which allowed the cyanide simply to trickle slowly through the mass of ore: in some circumstances this could replace the usual agitation with mechanical stirrers. Then, on 16 July 1888, the complete specification was lodged for the use of cyanide to extract gold from ores.[28] The Cassel Company had a process it could take to the gold producers of the world.

CHAPTER EIGHT

Transfer agents and colonial connections

The major gold producers of the world lay outside Europe. The cyanide process was a metropolitan development for use at the periphery, and its transfer to these countries was fundamental to its commercial exploitation. The method of transfer was therefore a vital policy issue for the Cassel Company. The cyanide process slipped readily into the system framework set up for the doomed Cassel method, yet accumulating experience would suggest several adjustments. For Australia, one thing remained constant—its position as a British colonial outpost provided many facilitating connections.

Transfer agents for Australia

The geographical scope of the Cassel system had been widely defined— Venezuela, Canada, Brazil, Costa Rica, Chile, Peru, Bolivia, Argentina, New Zealand and the Transvaal were all engaged in early negotiations— but North America and Australia topped the priority list, the potential there being considered 'vast', and 'practically unlimited'.[1] As early as September 1885, a five-year Australian agency was given to the British shipping firm of McIlwraith, McEacharn and Co. Not that others were not interested: there was indeed a spate of inquiries from Australian agents in late 1885 and early 1886. These included ore merchants, H. G. Lempriere and Co. in Melbourne, and Caird Paterson in Sydney. But the Cassel Board was convinced that their interests would be well guarded in McIlwraith, McEacharn's hands.[2] They had in Australia a multiplicity of personal, political and business networks, and a range of commercial and technical experience, which crossed both colonial and industrial boundaries.

The firm of McIlwraith, McEacharn and Co. had been formed in London in 1875 by two Scottish businessmen, Andrew McIlwraith and Malcolm Donald McEacharn. With family backgrounds in shipping,[3] and well-established contacts, the partners had soon won contracts to carry cargo and migrants to Queensland, a trade which was rapidly consolidated and extended with the help of well-connected relations in Australia. Andrew McIlwraith's brother Thomas[4] had been in the colony since 1854. His early days were spent as a railway surveyor and engineer, but his commercial instincts drew him into pastoral investments and a resolve to share in the meat-producing potential of Queensland and the meat market among British workers. Political fortunes had by 1874 made him Queensland Secretary for Public Works and Mines. Here he practised his philosophy of bringing migrants and capital from Britain to develop the colony, particularly through railway building, all of which provided lucrative government contracts for his brother's new shipping firm. Andrew McIlwraith's other brother John, manufacturer, Melbourne mayor and magistrate, was by 1876 the Melbourne agent for McIlwraith, McEacharn and Co., consolidating shipping interests by joining John Carson in a line of coastal coal and wool steamers.[5]

With the financial support of the Queensland National Bank, McIlwraith, McEacharn and Co. expanded their meat trade when the steamer *Strathleven* successfully landed the first cargo of frozen Australian meat in London in 1880. Their interests diversified when Malcolm McEacharn settled in Australia in 1881 and purchased for the firm the Rockhampton business of Walter Reid and Co., and also secured Queensland pastoral properties for himself and a share in a Mackay sugar mill. In July 1882, he married the daughter of Victorian mining magnate J. B. Watson. Further Victorian links were made as the firm began shipping coal under contract to the Melbourne Metropolitan Gas Co. The progressive shift from sail to steam brought new affiliations through interests acquired in the Queensland Steam Shipping Co. Here they came into contact with merchant James Burns, who in 1882 was planning to merge businesses with Robert Philp. He went to London to consult with Andrew McIlwraith and invite him to invest in the new firm of Burns Philp and Co. McIlwraith, McEacharn and Co. took an 8 per cent share, and Andrew McIlwraith became a director.[6] It was to prove a fortuitous relationship.

Developing transfer policy

From the outset it had been clear that the process would have to go to the ores, since the enormous cost of transport prevented their shipment to Glasgow for treatment. The cornerstones of transfer policy can be seen here: the conviction of the absolute necessity of conducting large-scale practical demonstrations to introduce the process, and the assumption that successful results would trigger rapid adoption, 'as a matter of course', throughout the country.[7] It was supported by a determination to send 'the very best men we have'.[8]

As Cassel shares rode high on the Glasgow Stock Exchange, Andrew McIlwraith organised the starting location for Australia. In 1885 he pointed out the attractiveness of northern Queensland and particularly Ravenswood and Charters Towers, with their 'unlimited quantity of valuable stone incapable of being treated by any Process'. Though initially he leaned toward Charters Towers, his firm in Rockhampton found a riverbank site near Ravenswood, roughly equidistant between Charters Towers and Townsville, with a railway close by. There, open sheds covered with galvanised corrugated roofing housed the Cassel plant sent out in 1886 and the engines, concentrators, gearing and tools obtained in Queensland.[9]

After lying idle under the care of Burns Philp in Townsville, the Ravenswood works were soon dusted off in 1888. The Cassel Board's chief object was to establish demonstration centres where the precious ores were abundant and sulphide ores had continued to defy a succession of methods. The directors' confidence in cyanide was boosted by 90 per cent extraction rates on two Ravenswood samples, though a third proved more difficult.[10]

Suitable staff were assessed for technical competence, practical mastery of the process, a conviction of its value, and enthusiasm for its successful exploitation.[11] Peter McIntyre was engaged to run the Ravenswood plant for three years, at a salary of £275 per annum plus travelling expenses. His brother Duncan would assist. As one of the Cassel Company's best laboratory workers, he received £150 per annum.[12] On 21 June 1888 the McIntyre brothers set sail for Queensland. A cable on 27 August signalled their safe arrival in Ravenswood.

The first task of the McIntyres was to show that the process would work, and at a viable cost. The local mining warden had already been sensitised to the potential of chemistry by a visiting lecturer from the Ballarat School of Mines and a local MP with experience in the research department of London's Royal School of Mines. He now believed mechanical means were 'entirely out of the question' and he was impressed with the McIntyres' early claims of extraction rates of 90–95 per cent, at a total cost of 30–40 shillings per ton. He was an eager agent of diffusion on his tours of the district and in reporting back to the Under-Secretary of Mines.[13]

The most suitable mechanism for commercialising the process was, however, still an open question. Back in the Cassel days, direct licensing from Britain to prospective South American users had included sale of apparatus and a fixed royalty on the process. For Australia, Andrew McIlwraith had suggested beginning with custom work, at £2 per ton.[14]

In 1888 this policy was resurrected, with a higher treatment charge, to maintain control of the process until patents were secure. The Cassel Board acceded to Peter McIntyre's request to scale up the plant, but urged caution until the patents were sealed.[15] In 1889 this plant became virtually self-supporting on the treatment of tailings for local mines.

During that year, transfer policy was in a state of flux. Patents were being confirmed in several countries and inquiries were coming in. A Melbourne firm of agents and importers, Campbell, Guthridge and Co., who also had an Adelaide branch, sought purchase of the patent rights for Victoria and South Australia. As 'a young and energetic firm' they were offered terms which required the formation of a company with a capital of £180,000 of which £80,000 would come to the Cassel Company, £45,000 in shares and £35,000 in cash.[16] Licensing options came under closer scrutiny when a Queensland mine-owner sought a licence direct from the Cassel Company and proposed himself as Cassel agent for his district. Though cautious, the Board thought they would impose a royalty of 7s 6d per ounce of extracted gold where a competing process was already operating, but seek a one-fifth share of capital or returns at new properties.[17]

Peter McIntyre's view that terms should be 'simple in character and direct in incidence'[18] gave food for thought as subsequent deliberations thrashed out some basic principles for licensing in the future. The first insisted that no expenses were to be incurred by the Cassel Company. The second set the royalty at 10–20 per cent of the pure gold extracted, with discretion down to 7.55 per cent as determined by the representative on the spot. The third defined three stipulations to attach to each contract: the Cassel Company would appoint a chemist to supervise the process, his salary to be paid by the licensee; the licensee would be obliged to obtain all cyanide from the Cassel Company during the whole term of the licence; and the licence would run at least for the whole duration of the patents and, if possible, for the full term of operation of the mine. Alternative licensing proposals, which would tie the Cassel Company into the risks of the user, were rejected as giving mine-owners undue power.[19] Control, guaranteed return and risk reduction were considered of greater value than potentially higher profit accompanied by high risk.

Though some moves were made to purchase mines or mining interests to work the process directly, the inclination toward the sale of patent rights also firmed up during 1889. Plans were on course for the sale of South African rights, and negotiations in Australia had shifted to sale of patents for the country as a whole. The Board also agreed with 'parties of high standing' to form a company to work the patents in the Crown Colonies of Hong Kong, the Straits Settlements, and North Borneo.[20]

By the time of the annual general meeting at the end of 1889, a clearer policy was emerging. The directors announced their intention to press for the formation of subsidiary companies to 'acquire and push the use of the process in other countries'.[21] The reasons were several, but all acknowledged the significance of system differences. Experience so far had shown that successful development of the business on a large scale in countries so far apart, and so distant from headquarters, could not be done efficiently by dealing with the individual properties. The local circumstances and the great variety of technical and commercial requirements could only be assessed on the spot. By putting the control in the hands of local parties, the company took advantage of their local knowledge, their direct interest in pushing the business, and their ability to decide all questions promptly. The mechanics of transfer to far-flung places were simply becoming too much for centralised control. This control would have to be mediated by the formation of locally based subsidiary companies, which could tune into the requirements of local technological systems.

'Bringing every influence to bear'

The need for local knowledge, sensitivities and networks was pressed home painfully in the case of Australia, where practical demonstrations and transfer were delayed by 'interference' with the Queensland patent.[22] Though the patent application was lodged and granted in December 1887, it was not finally sealed until nearly three years later, in August 1890. But if Australian conditions and events were responsible for the delay, Anglo-Australian connections ensured the eventual successful outcome.

Though the Cassel Board believed the Queensland patent difficulties were due to influences 'of an underhand character',[23] formally they arose from a prior patent application involving cyanide and changes in patent law. In 1883 six months' provisional protection, followed by six months' renewal, was granted for an application filed from Townsville in the name of Nicholas. The process described used cyanide with iodine, after roasting. Under existing patent law, no allowance was made for publication, or objection, but provisional specifications were open to inspection. By the time the provisional protection expired, however, the law had changed. The 1884 patent law, based on the 1882 British law, had set up a whole new system of patent procedures, including the provision for formal publication and objection, administered by an Examiner of Patents. When Nicholas applied for registration of a patent, the Examiner refused. But when the Macarthur–Forrest application was lodged three years later, two objectors said the 1883 Nicholas provisional certificate issued under the old statute was equivalent to letters patent under the Act of 1884, and the provision for inspection under the old Act was the equivalent of publication.[24]

When the Cassel Company received word of this obstruction in November 1888, it immediately began to organise its defences. The first tactic was legal and formal—an appeal to the Law Officer against the decision of the Examiner of Patents. In the case heard in January 1889, they were represented by counsel offering the authority of British scientific experts as evidence of novelty, and supported by the 'valuable help' of Robert Philp of Burns Philp and Co.[25]

When a reserved decision brought all the informal political and commercial connections into play, Robert Philp was even more helpful. Sir Thomas McIlwraith's National Party was in power and Robert Philp was himself now an MP, having entered politics in 1886 at the personal request and arrangement of Sir Thomas.[26] At the elections which brought McIlwraith to power in 1888, Philp changed seats to become the member for Townsville. From there, he would reach the mines portfolio by May 1893. Less obvious connections permeated a network of North Queensland representatives who were supporters of McIlwraith as well as active participants in the mining industry. In addition to Philp, there was James Macrossan, strong advocate for miners and mining, and Isidore Lissner, representative for Charters Towers. His local business had become a branch of Burns Philp and Co. in 1883.[27]

Philp and Andrew McIlwraith were also connected by business. Overlapping shipping interests of McIlwraith, McEacharn and Burns Philp[28] were compounded by numerous mining associations,[29] including a joint lease on the Rob Roy and Alexandria line of reef.[30] McIlwraith and McEacharn also acted for Burns Philp and Co. on the London stock exchange, their sale of shares in the Cumberland mines in 1886 realising a capital gain of £12,000 for the company in a year which was otherwise a poor one for Burns Philp.[31] Myriad crosslinking mining directorships added to the picture.

To help orchestrate the opportunities provided by these various colonial connections, Andrew McIlwraith departed for Queensland in March 1889. While cables received in Britain stressed that he 'was using every effort in order to arrive at a settlement',[32] the Board was also conferring about 'bringing every influence to bear',[33] with McIlwraith's Queensland partner, Robinson, who was in London at the time. There was also the option of using the Queensland Parliament as a forum for its case, 'if the other means now being taken should prove insufficient'.[34] This could be supported through connections into the British legal and political system, via the Attorney-General and the MP for Glasgow, Sir John Stirling

Maxwell. When Andrew McIlwraith returned to Glasgow, however, he advised the Board to hold back for the moment and to instruct the Ravenswood staff to take a low profile, attracting as little notice as possible for the present.[35]

In the meantime, negotiations were afoot to establish a syndicate to purchase the Australian patent rights. Prime movers were Wolston Trubshawe, a member of the South African syndicate, and Andrew McIlwraith. On 22 November 1889 McIlwraith conferred with the Board about the syndicate and Queensland patents. The two were integrally linked. The syndicate matters concerned the size and structure of the capital of the proposed Australian subsidiary. The patent matters concerned the use of some of the share capital, 'to be used if necessary for the conciliation of opposition in Queensland'.[36] On the first, a compromise was reached between the syndicate's desire for capital of £200,000 and the Board's desire for £300,000, by agreeing on £225,000. On the second, £25,000 of the capital would be reserved as shares for settling the Queensland patent matter and Andrew McIlwraith expressed himself hopeful of the removal of difficulties.[37]

It was Robert Philp, with the assistance of Lissner, who was the chief agent of this removal, through his political position and relations with the opponents of the patent.[38] Philp promised most of the 25,000 shares to those, including Lissner, who in some way assisted in securing the patent, thereby linking their interests to the successful flotation of the company, on which the deal was conditional.[39] The sealing of the patent on 12 August 1890 was followed a fortnight later by a letter sent from McIlwraith, McEacharn and Co. confirming the promise to send shares in return for the assistance provided by friends of Philp in achieving this end. Correspondence also continued with Philp, now a potential shareholder, on whether conditions favoured a float in Australia or England.[40]

In December 1890 the Cassel chairman reported to shareholders that the 'vexatious' difficulties of the Queensland patent, which had prevented any work being done in that colony, had been resolved 'through the joint efforts of ourselves and friends'.[41] The pivotal legal point was made by the Registrar of Patents, who revealed that recently received records showed that England's patent office had granted a patent to Macarthur and Forrest for the same invention under investigation in Queensland. The Registrar submitted that the applicants were therefore entitled to a Queensland patent. A later Chief Justice was to wonder at the relevance of the English patent in determining the novelty of the Queensland application, but the result was that the Law Officer overruled the objection of lack of novelty.[42] Colonial friends and British legal authority had won the day.

Forming an Anglo-Australian sub-system

With the patents finally secured, the Cassel Company could go ahead with the demonstration that was to precede formation of the planned Australian company. Burns Philp and friends would again oblige. Lissner and Philp arranged for the general manager of the famous Mount Morgan Gold Mining Company to provide both information and ore for treatment. Lissner went to Ravenswood to organise matters personally.[43] The trial conducted at Ravenswood early in 1891 processed five lots, each of 5 tons, of the most refractory ores, concentrates, and tailings. It was, 'as usual, a complete success',[44] yielding an average extraction of nearly 90 per cent, quite sufficient for the syndicate to proceed.

Although flotation of the Australian subsidiary was supposed to take place within six months of the demonstration, it was delayed well beyond this, again by Australian conditions. One factor was the sheer difficulty of reaching and persuading investors and miners when mining areas were so widely scattered over a large and isolated country.[45] Another was the onset of the 1890s depression, though its gravity was not initially recognised; a mild recession in Melbourne, resulting from the collapse of the speculative boom of land companies and 'land banks', quickly deepened and spread in the latter part of 1891.[46]

The first difficulty promised to be eased by taking into the syndicate an Adelaide merchant, George Swan Fowler, the principal partner in the firm of D. and J. Fowler and a past Treasurer of South Australia.[47] His firm, founded in 1854 as a retail grocery business, had by the 1880s established itself as a major trading company within Australia and, with a house in London, also between Australia and Britain. When George Fowler departed Adelaide in mid-1889 on a world tour, his keen interest in the future mineral wealth of South Australia kept him ever alert for processes which might deal with the colony's particularly refractory ores. A boating excursion with friends in England in early 1891 introduced him to the subject of the cyanide process and to the syndicate which was negotiating for South African patent rights. A visit to the Cassel works in Glasgow and a careful study of the company's business and process persuaded him of its potential. He agreed to join the syndicate floating the Australian company and to act as the medium for introducing the process to Australia. Completing his assessment with a diversion through the various operations in the Transvaal with John Macarthur, 'the more he saw the more he was convinced'.[48] By the time he reached Australia in August, the details of the agreement between the syndicate and the Cassel Company had been finalised verbally, and the formal document was about to be signed.[49]

The second difficulty was beyond anyone's control. The initial strategy was to float the company in Australia, a move Fowler favoured for two reasons. The Australian public should reap the financial rewards from Australian gold, he said, not British investors. And users would be more committed to the process if they could hold shares in the patentee company, all of which would enhance credibility.[50] As he travelled home, Fowler was expecting to float 65,000 or 70,000 shares without any trouble before the end of the year.[51] But he was not prepared for the rapidity of the decline in economic fortunes. On 20 August, as he noted the recent failure of two or three small banks, he was still 'sanguine' about floating in Australia. Two days later the shares of the Bank of South Australia plummeted from £25 to £15, causing great distress.[52] Fowler's letters from this time are full of the worsening crisis and an increasingly pessimistic prognosis for Australian flotation. As December brought news of the failure and suspension of banks and building societies, then associated 'frauds',[53] he lost all hope. As it became clear that all the Australian governments were 'hard up' and unable to float loans anywhere, he doubted whether the company could even be floated in London, since 'Australia and all things Australian do not seem very favourably received in England just now.'[54]

The Cassel Board responded to Fowler's gloom by agreeing to postpone the flotation from late January to late February 1892. This reprieve gave some relief but it tackled only part of the problem. On several occasions Fowler reiterated his view that the proposed capital of £225,000 was vastly unrealistic in the present climate. By December, he thought £100,000 a truer valuation. What he sought from the Cassel Company was an increase of working capital and a pro rata reduction in their number of shares. He argued that it was better to have 40,000 shares in a profitable paying concern, which was saleable, than to have double the amount in a company too heavily handicapped to do any good.[55]

The Cassel Board compromised by reducing the nominal capital to £180,000: £25,000 working capital; £20,000 cash and £50,000 in shares to the Cassel Company; a bonus of £15,000 for obtaining the Queensland patents; and shares worth £50,000 for the syndicate. The additional £20,000 would constitute a reserve. Fowler regarded the reduction of capital to £180,000 as a fair compromise, but not the retention of the 15,000 Queensland shares, especially when Cassel contributed only 2500 and the syndicate 12,500.[56] In fact these shares had always been a point of contention with Fowler, who felt that the current syndicate should not have to carry the burden of an agreement made by a syndicate of which he was not a member. It seemed as time went on that the rest of the syndicate came to share his feelings, and when Fowler returned to

Australia in mid-1891 he had the task of somehow trying to convince Philp to reduce his claim.[57]

When Philp offered little sign of accepting a reduced bundle of shares, or of even acknowledging Fowler's attempts, Fowler considered breaking away from the syndicate to renegotiate the patent purchase on his own behalf. But the matter was resolved more favourably than Fowler had supposed. Although in early 1893 he was adamant that 14,300 shares were the absolute maximum that should be allowed, by mid-1893 Philp had been persuaded to accept only 7500.[58] By this time he was Minister for Mines for Queensland, in a government headed by Sir Thomas McIlwraith, but his bargaining position had been reduced. Facing ruin from his speculative mining ventures, he was selling shares where and when he could.[59]

In the meantime, the Australian Gold Recovery Company (AGRC) was successfully floated in London at the end of February 1892. The final structure had given the Cassel Company 52,500 £1 shares as partial payment for the patent purchase at £135,000, out of the total capital of £180,000. Another 107,000 shares were offered for public subscription.[60] With almost a one-third share in the Australian company, the Cassel Company had two representatives on its Board—John Macarthur and Robert Smith.[61] Other directors linked the Australian company into the international network of Cassel subsidiaries. Charles McCulloch was chairman of the AGRC and its African equivalent. Edward Preston, Sir Charles W. F. Crawford and Wolston Trubshawe also served on both boards.[62] George Swan Fowler was the only Australian director.

The formation of the AGRC sealed the formal transfer of patent rights and the blueprint of the new technology. Experts in the chemistry and practice of the process had come some time before, together with some essential equipment; now came most of the capital. In putting together an Anglo-Australian sub-system, plans and expectations had been bent to the legal, economic and geographic environment of the colony, but colonial connections had eased the process. The result was two-tiered control from Britain, mediated by Fowler's Adelaide office, from which the Australian headquarters of the company initially operated. The AGRC was now the chief agent for diffusion of the cyanide process throughout the Australian colonies.

CHAPTER NINE

A challenge for technological imperialists

While British capitalists plotted and planned for the technological unlocking of the gold stores of Australia's ancient rocks, several indigenous factors combined to thwart their schemes and dreams. Despite the numerous economic and political links, assimilation of the cyanide process into local practice was no foregone conclusion. Though introduced to Australia in 1888 with the arrival of Peter and Duncan McIntyre at Ravenswood, its use was meagre until 1897.

Delays in adoption of new technology are not unusual, but Australia was notable for its tardiness in adopting cyanide: the Transvaal saw the process established as early as 1891, New Zealand by 1893–4, and America by 1895. In Australia, even a London-inspired special investigation into the AGRC operations in 1894 provided no solution to the lack of progress.

Some of the factors which held back adoption were built into the Australian system and were beyond the control of British capitalists. Others reflected a technological policy at odds with the conditions it confronted.

Facing the Australian system

Geography and geology were the first hurdles faced by the Cassel Company and its agents. Australia was isolated from the rest of the world, and its colonies were isolated from one another by distance, lack of infrastructure and political parochialism. Within colonies, vast empty miles separated mining districts scattered over undeveloped territory. In late 1891, the existence of the Ravenswood works was still 'practically unknown' on other Queensland fields. Even a curious Charters Towers miner would have to leave at 2.45 p.m. one day, remain all the next and return at 10.45 a.m. on the third day: 'it is not much to be wondered at', said George Fowler, 'if few of them visited the works'. For Fowler himself, it took at least nine days to make the trip from his Adelaide office, but he might still beat the post which took eleven days.[1] To visit more northern fields took sixteen days, while a trip to Yilgara in Western

Figure 9.1 Some of the significant goldfields for the Australian Gold Recovery Company in the 1890s

Australia required a similar period of travel in the opposite direction. He could not help contrasting this with South Africa, where operations were concentrated within a one-day drive.

South African ores had also seemed more compliant, readily releasing gold from reasonably uniform tailings. Out of the early confrontation with Australian ores, two kinds of difficulties emerged: either the extraction rate was not high enough to attract potential users, or so much cyanide was consumed, due to the presence of copper and other minerals, that the process was uneconomical to work.[2] Some samples exhibited both these problems in combination. Geography compounded these tricks of geology: the variability from colony to colony presented new technical problems on every front.

The relatively high cost of mining labour did not help the economics of the process. Rates varied from 7s 6d per day in New South Wales, to 8s 4d in Victoria, and 10s in Ravenswood. This compared with rates of 2s 6d to 3s 6d per day in the mining districts of Cornwall. Labour costs at 2–3s per ton of cyanided ore in Queensland compared with 10d per ton in South

Africa, where black labour was used.[3] At a time of economic depression, this only added to the lack of enthusiasm for risky expenditure on a new process.

It might be thought that the hazard of dealing with cyanide could itself act as a brake on the use of the process, but there was surprisingly little discussion of this factor, either with regard to the disposal of wastes or to the dangers for workers. Recommended precautions included the provision of long rubber gloves for workmen whose hands and arms might come in contact with cyanide solutions, and the ready availability of hydrogen peroxide as the most effective antidote. It was, however, pointed out that in countries like Australia and South Africa, where vats were not covered with roofs, the danger of inhalation of toxic vapours was 'practically unknown'.[4]

Instead, the Australian mining community as a whole was simply accused of prejudice against new methods, a product of the 'general conservatism of the old-time mining manager'.[5] Fowler formed the opinion that 'Australians are an unbelieving race on everything connected with mining',[6] and the *Australian Mining Standard* pointed out the anomaly of cyanide's great success in South Africa, while 'here the whole process is being received with mistrust'.[7]

Part of that response was rooted in the lack of chemical understanding and expertise on Australian mining fields. Blainey has argued that most mine managers in the 1880s were Cornishmen raised at a time when metallurgy was devalued by the abundance of cheap labour.[8] They had won promotion in Australia through their physical strength and their skill in handling men rather than ore. They respected practical experience over theoretical learning. Yet here was a process which required a hierarchy of technical skills—a chemist or metallurgist to determine 'by actual experiment' the exact concentrations and conditions for each kind of ore, a trained analytical chemist 'full of resource and originality' to adapt the process to daily variations in ore, a careful and trustworthy assayer to measure extraction, and an 'experienced and intelligent workman' with arithmetical ability to prepare and test working solutions.[9]

Cassel and AGRC policy reflected these requirements by including a 'competent chemist' in their licensing package.[10] But there were limits to the training that could be given to someone with only 'rudimentary knowledge', and long-term success depended on the quality of people available to work the process on a continuing basis.[11] Where were these people to come from? The *Australian Mining Standard* classified 80 per cent of alleged experts as 'pronounced duffers', incapable of treating refractory ores with any success.[12] A German engineer, Carl Wagemann, was amazed at 'how inferior the education of the men is here in mining'. Geologist Edward Dunn ventured that there were no more than half a dozen men in the colony competent to teach them.[13] As the major mining

institution, the Ballarat School of Mines had issued only thirty-six certificates of competency in assaying and metallurgy from its inception till
1888, only fifty-five by the end of 1891. Only twelve certificates had been
awarded in any form of chemistry. Even for these, Professor Alfred Mica
Smith admitted the training was limited, the period of attendance too
short.[14] Herein lay a basic problem of supply—a poorly developed technical infrastructure severely limited the means of producing the skills that
were needed and the mines were generally ill-equipped to assess and
harness the potential of such a process.

There were also very few people who saw the problem in the terms
which would bring this chemical process within their range of search
options. The cyanide process had been developed as a technical solution
to an economic problem—the demand for gold. But the economic
problem being experienced by the mining industry in Australia—declining gold yields and declining productivity—was experienced in different
ways within the Australian mining community. Some saw it as a lack of
goldfields, in which case the response was to seek new fields where
alluvial gold would bring quick and easy rewards. This was a strategy
used by governments as various means were devised for encouraging
prospectors to new areas.[15] Others looked to the discovery of new gold
bodies under the old fields, by means of deeper sinking. For those who
saw a failure of current techniques to obtain a payable return, the logical
response was to grapple with identified limitations. A quest for improvements in mechanical devices for crushing or concentrating ores was the
common line of approach. It emerged strongly from the Victorian Royal
Commission on the Decline in Goldmining, with the Government Analyst
being despatched overseas to look at new concentrating developments.

To the Government Analyst of New South Wales, chemical extraction
was the way of the future.[16] To those who had developed the cyanide
process, the equivalence of economic problem and chemical solution
was obvious. But even a special committee of the Australasian Association
for the Advancement of Science reported in January 1890 that this new
method seemed as yet impracticable.[17] To uneducated miners scattered
over the barren landscape of Australian goldfields, the connection would
not only be 'uncanny',[18] they would be lucky to even hear of it.

Negotiating the Australian system

The Cassel Company and then the AGRC were forced to accept certain facts of Australian mining life. Even where there was scope for ma-
noeuvre, the patentees were often slow to adjust. Part of the problem
was undoubtedly related to the fact that they had developed a transfer

policy on a global scale, and lacked either the resources, insight or flexibility to negotiate their way effectively through the idiosyncrasies of the Australian system.

If the British system was to make its mark, it would have to begin on Australian perceptions. The geography of Australian mining ruled out any comprehensive effort to win the hearts and minds of those who might use cyanide extraction. With the gradual recognition that 'what is being done in Queensland might not even be heard of in Victoria',[19] the patentees could only concentrate their efforts in those colonial centres which offered best prospects, and hope that good results would spread the word from there. Queensland, Victoria and South Australia took top priority. In each case the task was to target the most appropriate place to break down or slot into standard practice. Since practice and ores differed from colony to colony, this effectively meant working in three different directions at once.

From Ravenswood to Charters Towers

The hope of tapping into the wealth of Queensland fields[20] from a Ravenswood base seemed dashed by the withering effects of depression along with the failure to break down stubborn ores. Almost as stubborn was the standard practice of crushing with stamp mills, followed by amalgamation. True, there had been some toying with new techniques in the 1880s, but abandoned machinery stood as testimony to the tenacity of 'Ravenswood mundic',[21] and the McIntyres had not been able to pierce the descending gloom. Although in 1889 the plant had been paying its way, by 1891 Ravenswood mining had been declared almost 'dead' and Peter McIntyre was having trouble even obtaining tailings for treatment.[22]

Attention swung to the magic of Mount Morgan, about 400 miles to the south.[23] Its sheer wealth and high profile made it strategically important to the flotation of the Australian company, and it was currently engaged in systematic investigations under the force of General Manager Wesley Hall and the trained scientific hand of its metallurgist George A. Richard. Together they were pitting their experimental skills against the gold which was so fine it floated away, and against the coating of iron oxide which resisted amalgamation. Theirs was one of the few Australian companies consciously applying science to their problems and they were on a chemical track, having worked their way through several versions of the chlorination process. They were now forcing 95 per cent extraction rates from the Newbery–Vautin modification, developed by the Government Analyst in the Victorian Department of Mines and a visiting London metallurgist, Claude Vautin.[24] But they still faced the challenge of doing this at a satisfactory *cost*, especially on their difficult chalky kaolin ore.[25] Here was the point of attack.

There was, of course, the problem of royalty: Mount Morgan was paying no royalty on chlorination, while the Cassel levy of 7s 6d would clearly become prohibitive on ores which sometimes ran as high as 20 ounces of gold per ton. The Fowler–McIntyre strategy was to offer reduced royalty rates for very rich material and to highlight the simplicity and ease of the cyanide process against the complexities of the mountainside chlorination works. They also had another card up their sleeve in the experiments that Peter McIntyre had recently been making with the percolation style of cyaniding, which Fowler had seen operating in South Africa. By allowing the cyanide solution to seep through the ore, he had been able to avoid the impenetrable paste formed by agitating the fine and friable kaolin, a necessary step with chlorination.[26] In trials conducted at the mine in September 1891, pure kaolin responded to percolation with virtually 100 per cent extraction, for a soaking time of 24 hours and only 0.25 per cent consumption of cyanide. Fowler estimated costs would not exceed 9–12 shillings per ton, including cyanide and labour.[27] But it was not enough to win over Wesley Hall.

What would defeat cyanide at Mount Morgan was a combination of intellectual, capital and organisational commitment to chlorination, some personal pride and prejudice, the royalty, a rather narrow scope for penetrating current practice, and possibly some naivety in the design of trials. The intellectual commitment and prejudice came in the form of Hall's stated scepticism and warning that he had built his reputation on chlorination and they were 'not coming to destroy it'.[28] He interpreted the kaolin results to mean that at 70 hours to treat 6 hundredweight in a 5 foot 6 inch vat, output would be reduced from tons to hundredweights (i.e. to one-twentieth). He was apparently quite pleased to find that addition of the royalty to cost estimates gave warrant 'to debar all business'.[29] Fowler had to admit the naivety of trials targeting maximum extraction when cost was obviously the critical factor.

The breakthrough came instead at Charters Towers, the 'liveliest' Queensland field at this time, with shares booming on London markets. Methods had evolved around the easy winnings from the rich 'brownstone' deposits, whose oxidised quartz readily released its free metallic gold under the standard treatment of crushing and amalgamation. Characterised by its predominance of individual miners, the field was organised around central public mills where the ore was sent for crushing. Here the cost of milling had always been high, and efficiency low, but as brownstone gave way to poorer reefs and difficult sulphides no changes were made. The only modification was to collect the tailings for retreatment. These were concentrated and the heavier metallic portions ground with mercury in pans before further grinding and pan amalgamation. Costs could run to 35 shillings per ton, losses up to 50 per cent. The first round of fine grinding undertaken to release the gold from the ore compounded rather than cured the problem of loss through sliming, the formation of very fine particles which were carried away in the washing water.[30]

The first inroads of the cyanide method into Charters Towers were made with the sludges obtained after the fine grinding in pans. In late 1891 a trial of percolation brought the average treatment cost from 25s per ton by agitation to between 8s 9d and 9s 4d and eliminated the loss from the accompanying slime formation.[31] The sludges had been the mainstay of Ravenswood, but the future of the business could not be based on such a small part of the overall processing market. What they really needed for profitable working were the tailings concentrates, which represented 14 per cent of the total output of the field. If these were treated directly with cyanide, not only would there be 'a wide field' for the process, but the troublesome business of fine pan-grinding would be redundant.[32]

A trial designed to 'alter the current mode of work' achieved very favourable results on concentrates from two of the largest companies on the field. The price advantage of cyanide over fine grinding was £2 11s per ton for Day Dawn's concentrates and £4 2s 6d per ton for Golden Gate.[33] It was disappointing that the percolation which proved so economical on the sludges did not prove applicable to the concentrates, but since the result still left plenty of room for a 7s 6 per ounce royalty on 3–4 ounce concentrates, George Fowler predicted that 'this could revolutionise the mode of treatment in Charters Towers'.[34] He was sure that once one of the public mills adopted the process the others would be compelled to follow. The strategy seemed to be working when Thomas Buckland, manager of the English Day Dawn, but owner of the local Excelsior mill, signed up in early 1892 for the erection of his own cyanide plant, the first adoption of the cyanide process in Australia.[35] The AGRC would soon move its own operational headquarters from Ravenswood to this more promising site.

Daylesford failure

Though its position was being challenged by Queensland on annual production, in 1890 Victoria was still the leading gold colony, having yielded two-thirds of all the gold produced in Australia to that time.[36] It was still the centre of mining investment, enterprise and decision-making, and its government was actively trying to come to grips with declining gold yields through the technology-canvassing activity of its mines department and the evidence put to its royal commission on goldmining during 1889-91. It was important to have a nucleus of promotion here.

The opportunity arose through William Burrall, a mining and mechanical engineer with thirty years in the industry. He was an active seeker of better ways of treating refractory ores and had spent considerable time and money tracking down this new cyanide process. His search ended with the successful treatment at Ravenswood of his sample of pyritic North Cornish ore. He had passed on his view that this was the process of the future to the Victorian royal commision, but had been rebuffed by the experiments of the influential Government Analyst, James Cosmo Newbery, which were said to show the process was unsuited to fine gold. The cost of cyanide was said to make it commercially non-viable.[37]

There was a suspicion that Newbery was not altogether objective. While he had undergone a conversion from a thoroughly mechanical perspective on the problem to a perception of the chemical possibilities, he now had his own chlorination patent as a result of Mount Morgan investigations and had used his position to promote the Newbery–Vautin process throughout Victoria. By 1890, the pyrites works of Edwards and Deeble at Sandhurst, where much of Newbery's developmental work had been carried out, was using the process extensively on Victorian and other ores, and the Ballarat School of Mines, where Newbery was Examiner in Metallurgy, had an experimental plant for educational and testing purposes.[38] A disgusted Burrall 'had a very strong desire to expose the Minister of Mines and Mr Cosmo-Newbery' and was keen to start a small cyanide plant at Daylesford to carry out custom work for the surrounding district and provide 'an object lesson to the colonies'.[39]

Since the local ores contained only about 6 dwt of gold per ton, only concentrates would be feasible. As these ran at 5 oz or more, the royalty of 7s 6d per ounce again posed something of a problem, imposing a cost of 37s 6d before treatment costs of around 25s took the total to 52s 6d. The competing Newbery–Vautin chlorination charge was 60s per ton irrespective of gold content, but it was extracting 85 per cent. This meant that superior extraction would be the key to acceptance.[40]

As Mount Morgan faded from the picture, Fowler saw Daylesford as crucial to the successful float of the company. But when Duncan McIntyre arrived on 20 October 1891, he was confronted with a different set of circumstances from those he had been experiencing in Queensland. The demonstrated superiority of cyanide over fine grinding in Charters Towers offered little in Victoria, where fine grinding was not generally practised and concentrates were treated generally by chlorination. Instead he saw scope for percolation, as he found that 'the gold, *saved with the pyrites*, is exceedingly fine, and therefore more suitable for the quick action of the cyanide in a weak solution'. He predicted a 'striking future' for the AGRC, for Daylesford and all of Victoria.[41]

While Duncan McIntyre began to operate on samples of ores, Fowler set off on a series of visits to mining centres to stimulate interest in ore testing and investment in the Australian company. A trip to Ballarat and Melbourne picked up some promising samples and some useful contacts, including a meeting with the Minister for Mines. But, in a sign of the times, McEacharn's invitation to several mining men to meet Fowler at lunch fell flat when none showed up.[42] More alert to opportunity was the Hon. J. Wallace, 'a keen shrewd Scotchman' with interests in several mining ventures, and a serious seeker of useful overseas technology.[43] One of his most intractable problems was the abundant but low-grade and notoriously difficult ore of Bethanga, and he suggested that if the process performed he might even lend his influential name to assist the flotation of the company. In Bendigo, a meeting with 'quartz king' George Lansell and an introduction to mining magnate J. B. Watson indicated only slight prospects since the gold there was comparatively free, and the proportion of pyrites small.[44]

McIntyre's confrontation with ore samples from several mining districts of Victoria was even less heartening than Fowler's survey of investors. Only the Leviathan ore from Maryborough proved readily amenable to the process. All the others either gave poor extraction rates or consumed too much cyanide, or both. By late January 1892, Fowler had admitted 'Daylesford is not a success'. The process appeared a 'failure both as to pyrites concentrates and refractory ores'. For Fowler and Burrall this was a 'great blow', and they agreed that the best policy was 'reticence as to the actual results'.[45]

It was a pity, though, that time was not on their side. Persistent experiment with the cyanide-consuming Bethanga ores had made them amenable to percolation by early April. Duncan McIntyre would eventually find Victorian ores generally could be treated successfully if crushed finely to 60 mesh. But with minimal resources overstretched, Fowler had not the luxury to wait. Competition for the desirable concentrates was too keen; it seemed wiser to withdraw and focus the attack on more promising fields. He would rely instead on the Ballarat School of Mines to 'educate the coming men to the advantage of the process'.[46]

Home town advantages in Adelaide

George Fowler's home colony of South Australia was more receptive. As head of the mercantile business of D. and J. Fowler Ltd, he was an identity in Adelaide business circles. As a member of Parliament for the seat of East Adelaide from 1878 to 1884, and Treasurer in 1881, he knew his way around government circles.[47] These circles increasingly looked to gold to counter the colony's rising debt as the traditional staples of wheat, wool and copper suffered declining commodity prices from the 1870s.[48]

With known deposits of gold scattered over a large part of the colony, it was freely admitted that the South Australian problem lay in finding a suitable method to extract it from its ore. On returning from his world trip in mid-1891, Fowler combined advocacy of the cyanide process with his plans to stand for re-election in 1892. An interview with the *South Australian Register* made the process 'the subject of much talk and speculation' and opened up the lines of communication with mining activists awakened by the government's stimulatory measures.[49] One syndicate was keen to reopen the Mount Torrens mine and establish a cyanide plant to extract its gold. The government had agreed to contribute £500 toward mine development, and now approved its use for a cyanide plant on condition that the syndicate treat ore from any mine in South Australia, at not more than 35s per ton.[50]

To investigate this opportunity, Duncan McIntyre closed the Daylesford plant in February 1892 and departed for Adelaide. With the AGRC about to be floated in London, the pressure shifted to this new site. After the depressing failures at Daylesford, McIntyre was revitalised by consistently promising results, but the deepening economic crisis exerted its toll. First the Mount Torrens syndicate failed to float its company, then a revised plan fell through. The final blow came when the Federal Bank closed at the beginning of 1893; deprived of finance, all operations ceased.[51]

In the meantime, another South Australian mining authority had been in touch. David Rosewarne was a former Inspector of Mines and now manager of the Aclare and Westward Ho mines. He belonged to that small band of progressive mining men, as evidenced by his fellowship of the Geological Society and his appointment as a commissioner at the Mining and Metallurgical Exhibition in London in 1890.[52] He had contacted the Cassel Board while in Britain and now indicated his readiness 'to cooperate' if Fowler could demonstrate practical success.[53] Unfortunately, the high copper content of his Westward Ho samples consumed large amounts of cyanide, making the process uneconomical for their low levels of gold.[54]

Samples submitted from several other mines brought similar results: extraction was good and percolation worked well, but in most cases the gold content was too low for treatment to be profitable at current costs. Even on the rich 5 oz ironstone at Wadnaminga, cyanide seemed unlikely to compete with the current smelting cost of 50s per ounce when the royalty alone added up to 37s 6d. The Ajax ore also had higher gold content and looked hopeful for a time, but the closure of the Mount Torrens plant put an end to its prospects of treatment.[55]

Government channels opened up other prospects, helped by a brochure passed around parliamentary acquaintances. Several members of Parliament were interested to follow up the process with Fowler, some suggesting mines whose ores should be tested.[56] But of longer-term consequence was Fowler's arrangement with the Commissioner of Crown Lands to instruct the South Australian Inspector of Mines to include an inspection of the Ravenswood plant on his current investigative tour of the Northern Territory and Queensland.[57] Inspector J. V. Parkes visited Ravenswood late in 1891 and was favourably impressed. His report described the process as 'simple and efficacious' in the treatment of pyritic ores, especially those containing fine gold. Once again the royalty was the villain in the piece, condemned by Parkes in comparison with the Newbery-Vautin chlorination.[58]

But all was not lost. Conditioned by his government's determination to give a spur to goldmining, Parkes maintained a watching interest in the process and its viability for South Australian ores. The plans of the Mount Torrens syndicate had been undermined by financial crisis, but in October 1893, the member for West Torrens suggested the government should strike a blow for mining and employment by taking up where the syndicate had left off. The first step, he said, should be a report from the Inspector of Mines on the desirability of the government working its own cyanide plant.[59] By December, Parkes had made the case for the cyanide process, the virtues of a centralised cyanide plant, and the wisdom of using the now idle Mount Torrens works.[60]

Metropolitan policies

The lack of metallurgical and chemical skill at Australian mines might have been compensated for by exporting expertise as well as technology. This was indeed implied in the Cassel plan to train its own experts to take cyanide extraction technology to the world. In practice, however, the policy often fell down.

The AGRC needed its own supply of technical people to perform a range of activities: convincing demonstrations were imperative to persuade potential investors and users; visits to mines and mine managers were necessary to convince them to try the process on their ores; and adaptation of the process to different kinds of ore was vital to local acceptance. Yet the AGRC found that, as a subsidiary, it was not always easy to obtain the services of Glasgow experts. William Jones, the manager of the Cassel works, expressed a common feeling when he rejected a position in Western Australia because of the risk and the living conditions.[61]

From the outset Fowler was hampered by having 'only two experts', yet no help eventuated from his pleas for the despatch of competent operators.[62] Requests for demonstrations in other colonies received only a faint-hearted apology for the lack of chemists to respond.[63] Forced to concentrate resources for best effect, Fowler cast aside as impracticable, plans to leave Duncan McIntyre alone in Ravenswood, while he and Peter McIntyre visited other colonies to assess their potential.[64]

Even the experts at his disposal were a disappointment to Fowler. Peter McIntyre he judged 'competent but very slow'.[65] Plans for temporary use of a Cassel chemist returning from New Zealand to Britain via Australia were dashed when he proved unable to respond adequately to technical inquiries from Inspector Parkes.[66] Even the South African experience of John Macarthur's brother Charles did not make up for his lack of energy and self-reliance.[67] More to Fowler's liking was the combined business acumen and all-round mining knowledge of W. S. Paull, an Adelaide expatriate in Queensland, recruited as business manager by Fowler himself.[68] The most successful acquisition from Cassel was Gordon Wilson, a former chemist and assayer with Tharsis Copper Co. After introducing the cyanide process in Colorado in 1889, he had become managing chemist for the South African Gold Recovery Co. in 1891 and then manager of Charters Towers in 1893.[69] But another Cassel expert sent out in 1894 'would not set the Thames on fire'.[70] Curiously, the British directors of the AGRC were aware of staff limitations in Australia, but demonstrated little energy in facing this challenge, leaving Fowler in despair at the inefficiency of staff whom he had not chosen.[71]

The policy on royalty was also a frustration. Fowler was uneasy about its effect soon after his return home in 1891. The response of mining leaders, who described it as 'prohibitive', likely to 'kill any chance', and causing 'immediate abandonment', only added to his anxiety.[72] It was with envy that Fowler noted the ready payment of full royalty in South Africa, with an occasional 'impost' as well.[73] In Australia, 7s 6d per ounce on rich ores and concentrates virtually eliminated the cost advantage over chlorination, one of the motives for development of the process. With low-grade ores, treatment costs might be within the range of profitability but the addition of royalty could tip the scales into the area of unprofitability. In this case, the ores would stay in the ground.

The difficulty was clearly acknowledged by the Board in London, as it discussed the alternative of buying mines and ores for treatment. But licensing remained the prefered strategy, even though the royalty continued to be perceived as a barrier to adoption.[74] It was not until February 1896 that the AGRC finally succumbed to the weight of mining opinion and reduced its royalty to a basic rate of 5s per oz. The public rationale was that 'a lode yielding only 8 dwt of fine gold per ton by the amalgamation process alone would not pay the expense of mining, crushing, amalgamating, and developing without assistance'.[75] It was a change of heart pushed along by the pressure of developments on another front where the interaction of metropolitan policy and colonial will was already well advanced.

Patent uncertainties

Despite the patent obstruction in Queensland, most countries granted patent protection to the cyanide process without difficulty, affirming the Cassel Company's right to the rewards of discovery and forcing users of this new knowledge to return some share of the benefits. Then gradually a process of erosion of patentee rights and credibility set in.

The first signs appeared in the *Engineering and Mining Journal* of New York in January 1890, when a correspondent claimed that the use of potassium cyanide was neither new, nor effective.[76] The 1878 experiments of William Dixon of New South Wales were cited as part of the evidence. Further critical letters and comments over the next two years culminated in a devastating editorial of August 1892, heralding unequivocal evidence that the cyanide patents were invalid and the process unpatentable.[77]

Uncertainty has often been raised as a factor influencing decision-making about technology, and in late 1892 the infant AGRC was vulnerable as it started up its new plant in Charters Towers and prepared to bring the process forcefully before local miners. The 'crusade against the patentees' by the prestigious New York mining journal was reported swiftly in the *Australian Mining Standard*, but at this stage the journal sided with the *Mining and Scientific Press* of San Francisco, which was championing the cause of the patentees.[78] Then at the end of 1892 the South African Gold Recovery Co. was forced to take action against an infringer. In early 1893 the Cassel Company instituted proceedings in England against the Cyanide Gold Recovery Syndicate Ltd, for non-payment of royalties for the use of cyanide in its so-called Pielsticker process.[79] It was the beginning of what came to be known as the Pielsticker case.

Inactivity in Australia during 1893 was due as much to depression as anything else, but the climate of provocation was building up. Escalating reports of defaults on royalty payments in New Zealand and South Africa climaxed in the end-of-year news that a private Transvaal committee had concluded that the patents could be opposed successfully in a court of law.[80] By 1894, the process seemed poised for widespread adoption in Australian colonies: in Queensland, the AGRC headquarters was spreading its message from Charters Towers to other mining fields; in South Australia the government was erecting its own cyanide works as an impetus to others; in New South Wales, cyanide works had opened at Mitchell's Creek in the central west and negotiations were advancing with mining companies in Pambula in the south. But the events about to unfold would stall the whole process. As the royalty issue was shifted into new institutional domains, the courts of metropolis and periphery witnessed a struggle which saw out the century.

CHAPTER TEN

Governments, experts and institutional adjustment

British definitions of science and justice

The parameters of the legal struggle over cyanide were defined by British courts during the period 1893-95. The scene was set in the Pielsticker infringement case when Justice Romer interpreted the cyanide patent as a simple statement that *any* cyanide solution could be used to dissolve the gold.[1] Logic then told him his central probe must determine whether this fact was within the bounds of established chemical knowledge at the date of the patent.

The defending Pielsticker company was adamant that it was. Experts marshalled to verify this view included prominent consulting metallurgists Edward Riley and James MacTear as well as Professor John Attfield, all of whom testified that in 1887 no invention was needed to discover that a solution of potassium cyanide could be applied to dissolve gold in crushed ore. Opposing them were experts chosen from the cream of British science—Lord Kelvin, Professor W. Roberts Austin, Sir Henry Roscoe, Professor (later Sir) William Crookes.[2] They argued that 1887 science knew only that cyanide would dissolve *pure* gold. The Macarthur-Forrest process had then taken this chemical reaction and turned it into a commercial metallurgical process for extracting gold from *ore*. Justice Romer had many months during which to ponder the evidence, but the verdict finally delivered in November 1894 went to the defendant. Crucial to his assessment were earlier patents described by the defence experts, in which potassium cyanide was used to extract gold from its ores. The Rae patent of 1867 used potassium cyanide in combination with electric current, and the Simpson patent of 1885 used potassium cyanide in conjunction with ammonium carbonate, but the judge said that did not matter: it was known that potassium cyanide would dissolve gold.

The appeal case which followed, however, saw a shift in the central chemical issues. The defendants again argued lack of novelty and anticipation by prior patents, but a new emphasis touched interpretations as the Cassel case highlighted the commercial utility of the Macarthur-Forrest process, and the so-called 'selective action' of cyanide in leaching gold away from other minerals in naturally occurring ores. Before Romer, the Cassel counsel had called this selective action 'an accident', no part of the invention.[3] Now it was presented as the *essence* of the invention, and supporting evidence given 'unhesitatingly' by Professors Dewar, Austin and Crookes, among others, that this preference for gold over base metals was previously unknown.[4] Lord Justice Smith could see, then, that the problem the Macarthur–Forrest syndicate had defined for itself was how to extract the gold from the ore and into a state of solution without simultaneously taking the base metals known to occur in gold-bearing ore. He also accepted that the problem had been solved by the use of *dilute* solutions of potassium cyanide and that strong solutions would dissolve the base metals.

Influential in this assessment was the paper presented by William Adam Dixon to a meeting of the Royal Society of New South Wales in 1877, describing experiments conducted at the behest of the head of the New South Wales Department of Mines.[5] Among the twenty-four documents presented by the defence as anticipations of the Macarthur–Forrest patent, this paper was cited as the best example of prior knowledge and prior use of cyanide to extract the gold from ore. An opposing interpretation showed instead how it first dealt with the common knowledge that precipitated gold was soluble in potassium cyanide, but then presented a most significant statement:

> There being, therefore, no method by which the precious metals could be removed and the baser metals left, it remained to fall back on one of the first principles of metallurgy—viz., to remove the baser metals at the earliest stage if possible and leave the precious metals as a residue.[6]

If the selective extraction of gold from the base metals was the essence of the disputed invention, this statement suggested that Dixon was not aware of this characteristic of cyanide. Accordingly, Justice Smith took Dixon's paper as 'cogent evidence' in favour of the plaintiffs, for he saw that the prescription for first removing the base metals was exactly what the Macarthur–Forrest people had shown should *not* be done. Their invention was to remove the *precious* metals at the earliest stage, leaving the base metals as a residue.

As this became the determining factor in the case, Justice Smith found himself having to differentiate between the two claims made in the original patent. Since claim one appeared to apply to use of a known substance 'without stint or limitation', it not only deprived the public of a substance previously freely applicable, but most significantly, the judge had now been convinced that it had no utility: unless cyanide was used in the limited manner stated in the second claim, i.e. in dilute solution, it would dissolve the base metals conjointly with the gold, and give no beneficial result. On this line of argument, the judge was forced to rule the general cyanide claim invalid, yet the invention stated in the second claim was deemed to have novelty, invention and utility. On the technicality of claim one, the infringement action was lost, but the spirit of the decision was with the Cassel side. Justice Smith granted permission for the company to apply for a correcting amendment to the original patent, taking account of what he had now established as the crucial distinction between dilute and strong solutions of cyanide, and the selective action of dilute solutions. A revision of scientific fact had now gone into the legal records.

Australian repercussions

In Australia, legal assessments and verdicts were scrutinised for their implication. The regular reports of the *Australian Mining Standard* served as a major vehicle for communication of interpretations. But a decided change of tone had come upon the journal. The once-friendly advocate brought the news of Justice Romer's decision with a high-minded morality which declared that the patentees had grown rich on the enterprise of others, extracting 'a tribute almost as harsh as that imposed on the twelve tribes by Pharaoh', and probably with less justification.[7] But now that Romer had delivered the freedom to use cyanide, the journal could see progress ahead, including a stimulus to the research hitherto retarded by the 'irksome nature of the conditions imposed'. Australian mining companies were advised simply to wait out the finalisation of the legal process. The other alternative was to push for emulation of the South African example where, in order to thwart legal conflict, the state was said to be ready to acquire the cyanide monopoly and charge only 3.5 per cent royalty.[8]

Acknowledging to his son that it would now be harder to secure 10 per cent royalty in Australia, Fowler would have been pleased to negotiate with Australian governments, yet he knew that the coincidence of Romer's decision was more likely to make companies defer at least until the result of the Court of Appeal.[9] Even so reputable a body as the South Australian government was delaying payment of any royalty, and in Queensland there was general apprehension. The Charters Towers correspondent for the *Australian Mining Standard* threw an extra factor into the puzzle when it was observed that if Romer's decision were upheld, it would be useless to erect cyanide works unless agreements were entered with the AGRC for the use of zinc shavings.[10] By March 1895, the journal was deploring the effect of the uncertainty: 'in the meantime the industry flags, and money is being lost as well to the State as the individual'.[11] At Pambula in New South Wales they were certainly playing the waiting game. Negotiations suspended in 1892 had reopened in mid-1894, only to lapse again when the Romer decision came through.[12]

The import of that decision seemed confirmed by events in Germany. Berlin's Patent Office had originally accepted the cyanide patents, but in February 1895 the Imperial German Court followed Romer's line of thinking and pronounced them invalid.[13] Britain's Court of Appeal at first seemed to support this position by denying the Cassel Company its infringement claims against the Pielsticker process, and the *Australian Mining Standard* advised that no action could be taken by the patentees 'against any independent user of cyanide in these colonies'.[14] But, as the texts of the judgment became available, certainties evaporated once more. Even the Queensland Parliament was confused, and debated whether the British court had in fact ruled the patent invalid or not.[15]

Subsequent events did little to terminate the confusion or ease the strain on the AGRC. The much-publicised South African case[16] had been held in abeyance awaiting British outcomes, but now broke into a frenzy of international evidence-gathering by the protagonists. Added to the lag of uncertainty was the drain on the time, energy and resources of the Australian staff of the AGRC. The burden of organising Australasian material for the South African Gold Recovery Co. fell heavily upon James Fowler, George Fowler's son and deputy. Other staff were mobilised from around the country to present their own statements to the special commission which visited Australia in 1895.[17] Gordon Wilson took time out to conduct supporting experiments and to help solicit friendly testimony.[18]

At least in April 1895 Fowler could confidently state that in Australia there was 'no body of men banded together in opposition as in Transvaal'. But this was to change rapidly. On 1 May notice was received that three Queensland companies were about to cancel their agreements. Fowler considered 'it will be well to consider the matter from the point of view that we are now face to face with litigation in Australia'.[19] He was soon proved right.

From metropolis to periphery

The legal parameters had been defined in Britain, but the cyanide battle was now transferred to the periphery. To the Cassel Company, the way

had seemed to be cleared by the British Appeal Court's acknowledge-
ment of the novelty of the use of *dilute* cyanide solutions, and its leave
to seek amendment of the original patent specification. This was confirmed when the application for amendment was granted by the Comp-
troller of Patents on 20 August 1895, exactly as filed. Its patent affirmed,
the Cassel Company made applications for similar amendment in all the
goldmining districts of the world, including the Australasian colonies.

There, mining communities were caught napping, still puzzling
through the delayed interpretations of British decisions. Applications for
patent amendment slipped through unnoticed, validating the patent in
three colonies by the time local mining communities had caught up with
events. Instincts of vengeance were aroused; battle plans were drawn
up. They fell into two distinct campaigns, waged over a period of several
years. One was an economic campaign addressed to colonial governments. The other was essentially a scientific campaign, addressed to the
legal system.

The economic campaign

The economic campaign took off in the three colonies where the quick
pre-emptive strike by the patentees left an unprepared mining industry
without legal means for blocking royalty rights. Forced to seek other
weapons, mining representatives consistently advanced four economic
imperatives said to warrant government intervention on the side of local
mining:

- The cyanide process was essential for the extraction of gold from ores
  'which defy the ordinary methods of treatment', and for low-grade ores
  and tailings which could not tolerate high extraction costs.[20]
- It was impossible to work these materials profitably with the payment
  of the royalty, particularly the low-grade ores.[21]
- The treatment of these materials would provide considerable employment and a great boost to gold production, with benefit to all.[22]
- The excessive royalty not only inhibited production but also the
  research and experimentation that would extend the applicability of
  the method.[23]

The specified remedy varied in its content but not its objective—to
deliver the cyanide process to mining without the burden of royalties.
Colonial governments did, however, adopt their own style of response.

Queensland: pressure and principle

The political pressure in Queensland began as the real ambiguities of the new situation filtered through. Those companies which had boldly sent termination notices to the AGRC now sought protection from the government. First they wanted the Premier to grant financial assistance to resist the 'illegal' royalty, then they wanted the government to take its own action to challenge or revoke the patent.[24] Similar measures were pressed repeatedly in Parliament by goldfields representatives.[25]

But the Minister for Mines, Robert Philp, was not about to endorse a 'fishing career of litigation' against a patent validated by the courts of England, and held by a company in which he had shares. He and the Queensland government stood firm in defence of the status quo, the rights of the inventor, and the precedence of British law. While espousing the virtue of rewarding discovery, they decried the advocated evil of penalising a *particular* invention.[26]

Principle made concessions to pressure in 1896 by seeking details of royalty payments in other colonies. The AGRC was also canvassed on the possibility of some reduction, but the government pronounced itself well-satisfied with the defence of the royalty charges offered by Gordon Wilson.[27] Though miners subsequently extracted a face-to-face promise from the Premier to remove all such 'shackles' from mining,[28] any action was undetectable. As we shall see, this left the mining industry to take matters eventually into its own hands.

South Australia: a government buffer

In George Fowler's home colony, where the patentees also enjoyed some advantage, the government took on a different role. Having itself taken up the cyanide process as a means of stimulating a weak mining industry, the South Australian government came to the patent campaign as a cyanide supporter and user, as well as the people's representative. Its early cyanide advocacy had won it a royalty rate set at half the levy generally extracted,[29] but it was soon forced by parliamentary unease over Romer's decision to withhold the payments.[30] Fed by the urgings of mining deputations to defend the industry, this policy continued even after the appeal decision and was reinforced by advice in late July 1895 that the AGRC's claim for payment would probably be ruled invalid.[31]

Events had, however, moved on. Too late, a government inquiry to the Commissioner of Patents about revocation procedures and cyanide-related applications revealed that the amendment had been granted a month before. The AGRC's patent agent had appealed to the Commissioner's discretion to waive the usual public notification, on the grounds that the proposed alteration limited rather than extended the claims of the patentees, and would not affect the interests of others. A busy Commissioner complied.[32]

Recognising an irreversible situation, the government nevertheless continued to withhold royalties until threatened with legal action.[33] The mining community responded with pressure for the government to challenge the validity of the patents, annul the patent amendment, or purchase the patent rights. Though goldmining in South Australia was small scale compared with the major gold colonies, it was argued that the prospect of 'much mineral wealth' would be jeopardised by an economically unsustainable cyanide levy.[34] This presented a new dilemma. The government was keen to extend its policy of using its own cyanide works to fill the gap in private mining initiative, in the hope that this would stimulate clusters of adjacent mining activity. Yet all courses of action seemed to imply cost for the government.

An appeal was made to colonial unity in an attempt to rally a combined action by all Australian governments against the AGRC. But when this fell foul of colonial parochialism, the South Australian government opted for the role as public buffer for private benefit. Unattracted by the prospect of facing protracted legal disputation alone, it chose what seemed a far cheaper path in the end, and capitulated to the AGRC's claims for royalty in order that it might continue to sustain a fledgling industry.[35]

Western Australia: the legislative option

In Western Australia, an independent colony only since 1890, the patent amendment was merely notified to the local Registrar in 1895, without the knowledge of a mining community as yet untuned to the value of cyanide.[36] But by 1897, the production of cyanide gold was stepping up, and so was dissension against the royalty. By September that year, the local Coolgardie and Kalgoorlie Chambers of Mines had appointed a lobby committee. One of its tasks was to mobilise other chambers into a joint delegation to urge the Premier to annul the patents or to purchase the patent rights.[37]

Initially, the government refused to intervene on the grounds that the British court and its eminent scientific witnesses had established patent validity. When the 1898 Royal Commission on Mining canvassed industry opinion, however, thirty-eight witnesses and several written statements came down strongly in favour of government action.[38] At the same time the AGRC was aggressively pursuing the royalty arrears owed by Lake View Consols and warning other companies of the consequence of noncompliance. When the Chamber of Mines held its annual meeting at Kalgoorlie in November 1898, frustration was channelled into a plan to lobby the government and royal commission with the argument that the cyanide patents had been overruled elsewhere and should be tested as a national matter, at government expense.[39]

By now the proceedings against Lake View were in motion, but were long drawn out. The case was finally heard before a judge of the Supreme Court of Western Australia in September 1899, but had to be referred to the Full Court to decide certain legal points before the main issues of infringement and validity could be considered. As some of these went in favour of the plaintiffs and some in favour of the defendants, both took it on appeal and cross-appeal to the Privy Council. The latter gave a decision on the legal technicalities, leaving the infringement issues to be dealt with by the local court, except on the one important point of whether the specification had been duly amended in the colony. The Western Australian Supreme Court had answered this in the affirmative, in favour of the AGRC. The Privy Council reversed this decision on the grounds that the amendment had been *recorded* but not properly applied for and granted. Its reason was that any specifically *colonial* objection to the amendment had been given no forum.[40] This left the AGRC with an unamended, and therefore unvalidated, patent. The AGRC could neither enforce payment of past royalties, nor recover damages for past infringement.

However, it could apply anew for amendment, which it immediately did. The only way of avoiding future royalties, then, was to prevent the extension of the patents when they came up for renewal late in 1901. To this end great efforts were expended.[41] The result was special legislation to prevent an extension being granted in Western Australia. The Bill introduced by the Attorney-General on 11 October 1900 was rushed through by the end of November. Its purpose was stated as 'the regulation and controlling of applications for the extension of patents', but the House was informed that 'the particular patent which is, to some extent, practically the subject of this Bill, is the Macarthur–Forrest patent'.[42]

The *legislative* need for the bill arose from the relationship between Western Australian and British law. The Judicial Committee of the Privy Council could grant either an extension of the patent for seven years, or a fresh patent for fourteen years. A new patent would force the patentees to apply specifically for Western Australian registration, but an extension would simply flow on from Britain. The Bill was therefore designed 'so that, if the extension be granted by the Privy Council at home, it shall not be deemed to be an extension of the patent within the terms of our local Act, so as to compel us to register it here'.[43]

The economic motives impelling these legislative moves were clear in the Attorney-General's explanation that if the patentees were given another seven years, 'it will mean that this Parliament will have to . . . purchase the patent rights'.[44] And that was a very expensive option in Western Australia, where monthly gold production was running at around £500,000 per month, and half the gold of Kalgoorlie was said to be won by cyanide.[45] Moreover, thousands more men would be employed on the goldfields by ridding the industry of this 'great incubus'. Only the Hon. F. Whitcombe questioned the relevance of 'whether the dividend goes to the shareholders of a cyanide patent or the shareholders of different foreign-owned companies working in this colony'.[46]

Such quibbles were of no concern to the Kalgoorlie and Coolgardie Chambers of Mines as they informed the Legislative Council that the AGRC had acquired the patent for much less than amounts already received for royalties. The AGRC pushed the contrary petition that expiry of the patent in 1901 would give them only four years of royalty out of the term of fourteen, much of which was not even recoverable because of all the opposition.[47] But just reward for invention was not a factor as the completed legislation brought an end to Western Australian profits for the AGRC.

The scientific campaign

Events and methods took a different turn in Victoria and New South Wales, where unavoidable delays in the lodgement of the amendment application made secrecy impossible. Here science took the lead in the struggle to pre-empt the affirmation of royalty rights and protect the economics of local industry. Conducted against the backdrop of the issues analysed in the courts of Britain, the Transvaal and Germany, Australian proceedings were nevertheless guided by different parameters. Precipitated by the application to *amend* the specification, they focused neither on infringement nor on the validity of the original patent. Instead, they started with the fact that the original Australian specification made only *one* claim, the general claim for all solutions containing or yielding cyanide or cyanogen. The amendment therefore sought to disclaim the use of all strengths of solution and substitute the use of dilute solutions only. Under these circumstances, a shift in the emphasis of expert evidence was required on both sides.

Victoria: persistence, protection and purchase

In Victoria, an infringement action against W. M. King of the Golden Mountain mine delayed the amendment case until May 1896. The Commissioner of Patents was then asked to allow insertion of the words 'a dilute solution' before 'cyanogen or a cyanide or other substance or compound containing or yielding cyanogen'. The objection was raised that this would describe an invention substantially *different* from the original. The Macarthur–Forrest side wanted to establish the opposite. The result hung on the distinction that could be made between dilute and strong solutions.[48]

For the AGRC, Britain's Professor Dewar submitted that scientific orthodoxy took dilute solutions of salts to be those in which the physical characteristics of the solutions were not materially different from the same characteristics of the solvent liquid. He offered a strength of 5 parts in 100 as a general cut-off point. Australia's C. R. Blackett, Metallurgy Examiner at Ballarat School of Mines, concurred, and supporting declarations from G. A. Goyder and Professor Orme Masson added weight to their case. The opposition denied this orthodoxy and retorted that without specified dilutions there was insufficient detail to enable immediate working of the process, a significant point in a colony whose patent law had always required clear specification of 'the manner and process of making constructing using and compounding' the invention.[49] On the witness stand, cyanide researcher Dr John Storer swore that the amendment would not allow implementation of the process without troublesome and expensive experiments to determine the appropriate strength of solution. Supporting declarations came from William Dixon, G. S. Duncan, and New Zealanders H. A. Gordon (Government Metallurgist), William Skey (Government Analyst) and James Park (Director of the Thames School of Mines).

The Commissioner came down on the side of precision and ruled that concentrations should be specified to allow people of ordinary skill and aptitude to work the process without 'troublesome and expensive experiments'. The amendment, he said, did not disclose any such 'exact proportions' as the mining community had a right to expect, while its timing suggested the introduction of subsequently acquired knowledge: 'if an omission occurred through inadvertence, an amendment should have been sought promptly, and delay for upwards of seven years calls for an explanation'. Commissioner Akehurst admitted his decision had weighed the 'considerable importance' of this case to 'the mining public of Victoria'.[50] It also emerged that the government had in fact been secretly backing objectors, spurred by parliamentary pressure to protect the mining community.[51]

The Victorian amendment was refused in June 1896. Soon after, the long-awaited South African case reached court, with testimony gathered from around the world. By this time, the selective action of dilute cyanide solutions had become a central issue for determination and the subject of much scientific experiment.[52] The Transvaal judge was prepared to accept experimental evidence that selective action did exist, but his chief concern was *who discovered it*. Expert witnesses argued that the mere dilution of a chemical reagent was no discovery but the inevitable result of simple economy in the use of expensive chemicals. Australia's Dr John Storer added that the moment a chemical reaction became public property anyone was free to make any strength of solution, and that there was no novelty entailed in the reference to certain proportions of cyanide. The final judgment ruled that any selective action which did exist had not been discovered by the Macarthur–Forrest syndicate, and was in any case not a discrete invention but an unpatentable chemical law of nature.[53] On 4 November 1896, the South African patent was cancelled.

Back in Australia the whole matter was still up in the air. In Victoria, the AGRC had, on 25 June 1896, submitted a new application adjusted to accord with the previous ruling. Its more specific claim was restricted to the use of a solution not exceeding 8 parts of cyanogen to 1000 parts of water. To forestall allegations of subsequently acquired knowledge, John Macarthur declared that the 'working tests' needed to scale up to industrial operations were not 'of a kind to yield novel results capable of constituting patentable invention'.[54] This sounded remarkably like the arguments for the *opposition* in South Africa, but he would still be checked by the opposition in Australia. Since the amended British patent claimed dilutions between 2 and 8 parts of cyanogen per 1000 parts of water, the new Victorian claim for *all* dilutions below 8 parts per 1000 was said to be a *larger* claim than the original and than that which had been validated in Britain. It must therefore be based on subsequently acquired knowledge. In early July 1897, after a hearing lasting ten days, the new Commissioner for Patents confirmed the decision of his predecessor and refused to allow the amendment.[55]

When an appeal to the Attorney-General reaffirmed the decision, the AGRC went to the Supreme Court, to be heard by Justice Holroyd, who agreed to take fresh evidence. Again he found the amendment to be substantially different from the original claim. In dismissing the patentees' case, Holroyd declared it 'almost incredible that if he was cognisant of the selective action of dilute solutions, and desired to claim it, he should have wrapped up his knowledge in the vague generality . . . of the Victorian specification'.[56] From the decision of the Supreme Court the AGRC took an appeal to the Full Court, finally winning on the basis of the fresh evidence submitted earlier.[57]

The fresh evidence was accepted as necessary to reveal the state of knowledge of the persons for whom the specification was intended, namely competent and willing chemists. Consequently, the state of chemical knowledge in 1887–88 once again became the pivotal issue.

On the AGRC side Professor Masson and other experts were emphatic that a competent man reading the specification in 1888 without prejudice would find several pointers to the use of weak solutions in existing knowledge and textbooks. Economy would also urge weak solutions when all the other guides pointed the same way. On the other hand, opposition experts testified that such a man would at once carry out his own research to determine if the textbooks and current knowledge were accurate, and that he would use medium-strength solutions. Since the patentees had failed for so long to discover the advantage of weak solutions, an ordinary investigator would be at least equally perplexed.

Justice Holroyd had taken a similar view. But conflicting expert evidence split the Full Court. Justice Williams looked back to the previous rulings in Britain and Victoria that the *general* claim was of no utility, and asked: 'How, then, can we allow experts to come forward and say—"The Court of Appeal in England is wrong, and Holroyd is wrong?"'[58] But Justices Hood and A'Beckett accepted that the patentees had from the beginning disclosed an invention which would 'disentangle' the gold from its ores, that their use of dilute solutions followed from the chemical knowledge of the day, and that later disclosure of more precise concentrations was merely a statement of the particular weak solution which, as a rule, would be most advantageous. With two to one on the side of the AGRC, 1899 saw their amendment finally allowed.

The chemical arguments of the AGRC's experts had finally triumphed, but economic imperatives urged the government to take another course to free industry from the royalty. The *Australian Mining Standard* predicted that the judgment would handicap all cyaniding operations, especially at Stawell, Bendigo, Tarnagulla, and Egerton, where the process was in use on a considerable scale.[59] But as the AGRC began action to claim its dues, W. B. Gray was already appealing to the Privy Council, on behalf of all objectors, against the Full Court's decision. In the process, it became clear that Gray had all along been backed by the Victorian Mines Department, and that support was continuing still to Gray and others facing infringement action. In public, the Minister and the Premier claimed their action to protect the industry was justified by the significance of the royalty for the balance between profit and loss on low-grade material.[60]

Having spent close to £5000 already, the government was now pressed to purchase the patent rights. Parliamentary representatives pointed to paralysed mining districts and suggested that as patentee the government should set a royalty that would just recoup expenditure. All advice indicated that this might be the least painful way to resolve the issue. As a businessman, the Premier's instinct was to strike a bargain before the final decision of the Privy Council, and to feel out the minimum accept-able price by discreet private channels, but he was adamant that the AGRC was not going to dictate terms.[61]

The *Australian Mining Standard* reminded the Premier of estimates that the process had added £200,000 to the annual gold yield of the colony, and given employment to 1200 men. It was clear the general welfare required that its use should be freed from all restrictions. When the final purchase figure of £20,000 was announced at the end of January 1900, the Minister for Mines, the *Australian Mining Standard* and the industry were well pleased.[62] For a sum which the Minister admitted 'had often been frittered away in matters of very much less importance', a process had been acquired which he estimated would employ another 3000–4000 men.[63]

The enabling Bill provided for the government to charge only 2.5 per cent for a cyanide licence, payable monthly. To enable recovery of the purchase cost, the legislation would extend the term of the patent to five years.[64] From 1905 the use of the cyanide process in Victoria was free. The *Australian Mining Standard* rejoiced that this most effective of all gold-saving aids 'may henceforth be fully used without challenge'.[65]

New South Wales: patent reform versus 'foreign exploitation'

Conditions in New South Wales led to yet another form of institutional adjustment to the royalty. Here no amendment was possible until after passage of the *Patents Law Amendment Act* late in 1896. This legislation itself generated considerable heat, it being supposed that it was being passed merely to allow the AGRC to amend its patent.[66] In fact the moti-vation was to bring New South Wales law in line with the English Act and to expedite the working of useful patents by allowing certain changes in specifications. At the time of introducing the Bill the Hon. J. H. Want had never heard of the AGRC and 'the attempts to levy blackmail'. But by late 1896, a stormy reception in the Legislative Assembly had dispelled his ignorance.[67] The Hon. J. Creed focused on what was then the nub of the problem for the government—how to provide the conditions to help other worthy patentees, while maintaining protection against the likes of 'the cyanide corporation', which he described as:

> one of the richest corporations in the world, containing men of position who are able to influence foreign governments and men who are willing if neces-sary, to spend hundreds of thousands, possibly millions of money, in carrying out their ideas in a particular way.[68]

The legislative compromise was found in cancelling the retrospective aspects of the Bill which might injure people who had invested their money in works under the existing state of the law.[69]

The Act was finally passed on 16 November 1896, and on 13 January 1897 the documents for the application to amend the cyanide patent were signed in London. They were filed in the New South Wales Patent Office at the end of February.[70] Opposition was immediately organised on a wide scale. Newspaper notices provided details of the proposed amendment, and George Massey, consulting mining engineer, prepared further information for a public meeting to be held in Sydney on 20 March to organise action to contest the amendment. An impressive list of mining industry representatives formed the New South Wales Cyanide Opposition Committee for the purpose.[71]

Among the committee members were active opponents of the 1896 Patent Amendment Bill. Several of the mining companies they represented were also among the formal objectors to the cyanide amendment, most of whom joined forces with the one counsel, organised by the Opposition Committee. Other formal objectors were John Deeble from Victoria and Duncan Noyes and Co., both of whom had made improvements to the cyanide process since its introduction into Australia. From the government, Attorney-General J. H. Want and Minister for Mines Sydney Smith formally lodged their notice of opposition on 12 April 1897.[72] In this mixed alliance against the AGRC, government concern for the generation of public wealth met with more specific interests, both local and foreign, both cyanide user and competitor.

When a lone voice writing under the name 'J' criticised the intervention of government ministers, Arthur Griffiths, MP and patent agent, countered that the patentees 'claimed what they did *not* invent' and that the amendment would affect 'the treatment of the whole of our low-grade ores'.[73] This 'duty' to protect mining and public interests was pressed in Parliament, and also gained support from within the Royal Society of New South Wales when President Henry Deane, in his address of 1898, commended the government's action, which 'in resisting the attempt to create a cyanide monopoly in this colony has resulted in the successful introduction of the process on many of the New South Wales goldfields'.[74]

The filing of evidence went on for more than eighteen months in New South Wales. As one side brought forward more scientific, expert evidence, the other side brought their own expert witness in reply. Scientists *supporting* the amendment were Professor Archibald Liversidge (Chemistry and Geology at Sydney University and head of its School of Mines), Professor Alfred Mica Smith (Ballarat School of Mines), and Professor Black (Dunedin). Supporting chemists and metallurgists using the process included G. A. Goyder from South Australia and also patentee employees like Gordon Wilson of Charters Towers and Howard Greenway of the Cassel Company in New Zealand. Expert witnesses *against* the amendment were W. A. Dixon (analytical chemist, teacher at Sydney Technical College, and author of the much-quoted 1877 paper on the use of cyanide), J. Walker (chemist to Bar-medman mining company), J. C. H. Mingaye (NSW Government Analyst), James Taylor (NSW Government Metallurgist), James Park (Thames School of Mines, New Zealand), and William Skey (New Zealand Government Analyst).[75]

The hearing finally came before the Examiner of Patents on 13 December 1898. Here, where patent amendment had until so recently been impossible, another variable had been added to the legal parameters. In particular, much was made of the capital already committed to cyanide works on the basis of the combined effect of the British appeal decision and prior New South Wales patent law. A Victorian user, John Deeble, argued that as a consequence of advice that patent 453 was invalid and could not be amended he had made arrangements to build cyanide plants in New South Wales and to treat ores with extremely weak solutions of cyanide.[76] Once again a central issue was the relation between the use of strong versus dilute solutions. Once again expert witness seemed to shift emphasis to suit a new context, as an impressive array of scientific personnel argued their side's case with a thrust opposite to that adopted in South Africa and the British Appeal Court.

In addition, a new chemical issue had been added to the agenda when the New Zealand Government Analyst, William Skey, challenged the general assumption that cyanogen did in fact dissolve gold. Experiments reported in 1897 showed that gold-leaf subjected to cyanogen showed no sign of dissolving even after several days. Skey argued that whereas alkaline substances had the effect of decomposing cyanogen into *products* that were solvents of gold, in practical cyaniding the acidic reaction of air, water, and quartz reefs would prevent decomposition. The patent was therefore wrong to claim cyanogen as an alternative to alkaline potassium cyanide, the latter being the *only* solvent.[77]

Skey's statement to this effect set off another round of declarations. One set of arguments dealt with the semantics of the specification. Professor Alfred Mica Smith said the specification was for a solution *containing* cyanogen, not for *pure* cyanogen in pure water. Any solution in which the radical cyanogen existed, either free or in combination, met this requirement and since the radical cyanogen was the active agent, it was fitting to specify exact concentrations in terms of cyanogen. John Mingaye, New South Wales Government Analyst, sided with Skey's case that the action of cyanogen on some ores in the presence of alkalis could not justify a general statement that cyanogen was capable of extracting

Figure 10.1 Howard H. Greenway, formerly of the Cassel Company, New Zealand, was now general manager of the Australian Gold Recovery Company (*Australian Mining Standard*, 11 May 1899)

gold from ores. Cyanogen did not have this power, though certain compounds containing cyanogen did.

Another set of arguments was based on actual experiment. Howard Greenway found the use of solutions containing cyanogen gas satisfactory on both powdered ore and pure gold. Furthermore, the solvent power of cyanogen gas solutions was equal to or better than that of potassium cyanide solutions. Dixon said his technique must have been faulty—since the cyanide of gold was practically insoluble in water unless alkaline cyanides were present, ammonium cyanide and not cyanogen must have been produced in Greenway's retort. The New South Wales Government Metallurgist, James Taylor, had come to a similar conclusion after repeating Greenway's experiments.

After a lengthy hearing, the Registrar of Patents delivered his judgment on 19 June 1899, and refused the amendment. The AGRC announced its intention of appealing, but since the 1896 Patent Amendment Act had put the decision in the hands of the Minister for Justice, rejection of the amendment was confirmed.[78] Another colony had escaped royalty payment.

Queensland again: arguments old and new

There was one last patent case to be fought out in Australia. This was in Queensland, where the government had resisted the pressure of economic argument to move against the AGRC. But in 1900, with patent renewal possible the following year, the Day Dawn PC Gold Mining Co. Ltd precipitated legal action by refusing to pay any more royalty. Instructions from London headquarters told the local manager to rally general mining support by canvassing opinion among mine-owners and cyanide workers.[79] The AGRC responded with proceedings against Day Dawn, and on 17 April 1902 the hearing began in the Supreme Court in Brisbane before the Chief Justice, Sir Samuel Griffith. Griffith had been judged a friend by the Cassel Company when pursuing the original Queensland patent,[80] and was the coalition partner of Thomas McIlwraith when the original patent was finally granted. Now he alone was to judge the merits of the case.[81]

The amendment having gone through in Queensland without query in 1895, the issue now was a test of validity of the specification for which royalty was being sought. While the chemical issues were familiar, there were new twists in the use of scientific evidence to suit the specific conditions. One particularly Queensland element was the inclusion among anticipations of the specification granted to Nicholas in 1884: it was this which had caused so much trouble and delay in the granting of the original patent. Macarthur's directions to divert South African Commissioners from Brisbane so as to avoid 'the unearthing' of this specification[82] had apparently been effective to date, but the Nicholas process would now be fully aired. Since this used cyanide plus iodine to dissolve gold from roasted ores, it posed the question of whether Nicholas' specification was for a cyanide process or an iodine process. Howard Greenway, for the AGRC, insisted it was an iodine process. He was supported by John Brownlie Henderson, Queensland Government Analyst and formerly Research Assistant to Professor Dittmar, who had testified for the Cassel Company in Britain. By contrast, William Dixon testified that experiment had led him to conclude that the active agent was the cyanide, the iodine simply hastening the reaction. Day Dawn counsel followed up with the assertion that the iodine merely substituted for oxygen in accelerating the reaction. Henderson countered that outside laboratory work, iodine was never used as an 'oxidiser'.[83]

Also new was the centrality of the recently questioned *solubility* of gold in cyanogen. Against the AGRC's affirmative evidence, William Dixon testified that many cyanogen-yielding substances would not transfer their cyanogen to the gold, and cited his own experiments showing that mercuric cyanide would not act as a solvent, and that cyanogen gas would not dissolve gold. An analytical chemist, Robert Mar, testified likewise, saying the bromide, chloride and iodide of cyanogen all contained and yielded cyanogen, but did not dissolve gold. The defence counsel rounded off the argument by declaring that any competent and willing man who followed the specification in 1887, and used cyanogen gas, would therefore be misled. The fact that any ammonium cyanide present would assist the process was accidental and would be cancelled where mineral acids were present.

As for selective action, in this patent validity case the opposition had returned to a position of playing down both its existence and its 'discovery' by the Macarthur–Forrest syndicate. And now William Dixon could add testimony based on his recent research designed to show that there was no such thing. Any apparent selective action, he insisted, was in fact dependent on the surface area of *any* metal exposed, and arose from the action of *oxygen* present in dilute cyanide solutions making possible what could not be achieved by cyanide alone in strong solutions.[84]

On the other hand, the AGRC counsel had reverted to the opposite position. Greenway pointed out that the idea of selective action had at first been ridiculed, and insisted that the state of knowledge before 1887 would lead away from rather than toward the cyanide process, because the base metals were also affected. Duncan McIntyre asserted that if base metals were present in large quantities, the selective action of weak solutions was conspicuous. George Ernest Bray corroborated this and Gordon Wilson backed up the statement that selective action was unknown before 1887. Once again the emphasis of each side had changed according to the relevant legal parameters. This time, Queensland's Chief Justice came down clearly and unequivocally on the side of the AGRC.

In the determination to have the cyanide process on minimum terms, economic and scientific arguments had been advanced by an alliance of mixed interests, including those who simply wished to use the process as cheaply as possible, those who wished to promote its use to augment the production of public wealth, those who had competing methods, and those whose incremental improvements solved particular locational problems. Harnessed to·the rhetoric of a growing Australian nationalism, their claims were united in a largely successful way. With the help of governments and experts, the cyanide process was used in Australia without payment to the AGRC of the enormous sums originally anticipated. Only in Queensland, where the Cassel Company had such good connections, was the industry left to incur the costs without aid. By now, though, they were finding it worth their while to pay the price.

CHAPTER ELEVEN

From Glasgow to Kalgoorlie

Urgent pleas to Glasgow to increase cyanide shipments from February 1897[1] signalled a change of tide in the fortunes of the cyanide process. Australian gold production by cyanide would soon become significant on an international scale. By 1900 it had surpassed that of the Transvaal, and at 683,900 bullion ounces was well ahead of the USA at 497,280 and New Zealand at 452,524.[2]

Overall, the number of Australian users paralleled the amount of gold recovered by cyanide, jumping remarkably in 1897 and increasing over the next few years. Charters Towers led the way, pushing Queensland to its peak by 1899. Western Australia soared to maximum gold production and cyanide consumption in 1903. The number of cyanide plants peaked ten years later, about the same time as in Victoria and New South Wales. South Australian trends were similar. By all indicators, a plateau was reached during the early years of the twentieth century before the cyanide process succumbed to another downturn in mining which was overtaking the new states of Australia by 1910.

Colonial patterns of adoption

Queensland had been nominated by the Cassel Company as the original nucleus for diffusion of cyanide technology in Australia and it was here the process first became an important addition to extraction techniques. At the new AGRC centre in Charters Towers, expenses could be covered by treating concentrates, tailings and sludges for surrounding mines. But more important was the opportunity to exhibit the merits of the process and provide the trial facilities which would reduce the risk and uncertainty of investment. From this thriving centre the process was actively carried by AGRC chemists to widely dispersed goldfields.

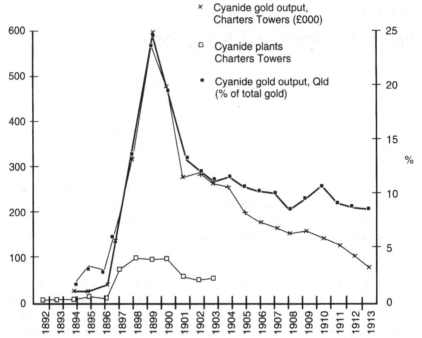

Figure 11.1   Cyanide gold production, Charters Towers and Queensland, 1892-1913
Source: Annual Reports, Queensland Department of Mines

First in were the large mining companies, like Day Dawn and the Victory at Charters Towers, then the Ravenswood Gold Company at Ravenswood and the Quartz Crushing Co. at Croydon. The cyanide plant which opened there in mid-1894 cost £6500 and was the largest of its kind in Australia.[3] But in Queensland the cyanide process took off with the small producers. In 1896 the AGRC put up twenty-five new plants in Queensland, many of them dealing with only 200-250 tons per month.[4] By 1897 the cyanide process had really arrived, especially on the Charters Towers field, where miners were said to be 'cyanide mad'.[5] Here, many individuals or small collectives were putting up small plants for £100-200 to treat the discarded tailings in the water courses. In some cases there were five to a paddock.[6]

In 1898 it was pointed out that Queensland was unique, cyaniding having been taken up 'as a poor man's process'.[7] In this respect it reflected a pattern of investment where nearly 80 per cent of Australian goldmining capital came from Britain, but less than 5 per cent of this went to Queensland. Queensland goldmining was a home industry, with 95 per cent of its funds provided by local residents.[8]

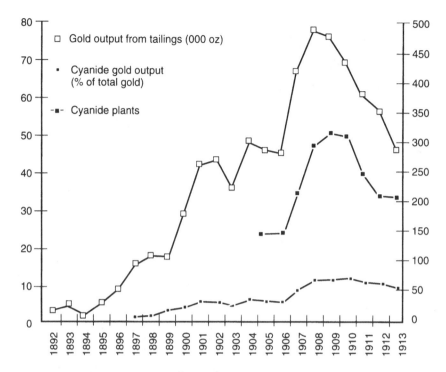

Figure 11.2 Cyanide gold production, Victoria, 1892-1913
Source: Annual Reports, Victorian Department of Mines

In 1899 cyanide recovered 22 per cent of the total gold production for the colony. Charters Towers still dominated, gaining 40 per cent of its own gold value by cyanide and contributing 89 per cent of Queensland's cyanide gold.[9] Chlorination, smelting and alluvial mining were by this time contributing only 2 per cent.[10] In 1900 the manager of one of the mines assessed the average benefit of the cyanide process to the field at an addition of 4 shillings to the value of each ton of ore produced, providing a net additional profit of £48,000 per annum.[11]

In South Australia, a much weaker goldmining industry was held back not by lack of gold but by its dispersed distribution and its difficulty of treatment. Pitted against these quirks of nature was a co-operative relationship between the government and the private mining community. The result was two parallel and complementary centres for diffusion, one private, the other a government enterprise designed to fill the void of private initiative.

The private centre for cyanide treatment emerged around the Virginia mine at Wadnaminga, whose success spurred the releasing of many abandoned mines in the once rich alluvial area between Manna Hill and

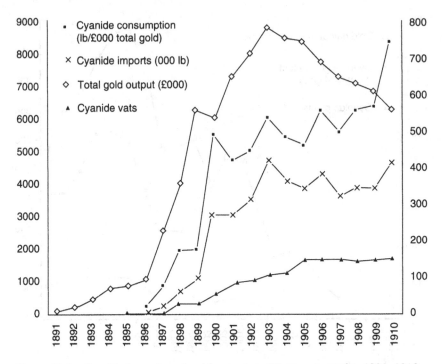

Figure 11.3   Cyanide imports and gold outputs in Western Australia, 1891–1910
Sources: Annual Reports, WA Department of Mines; WA *Yearbook*; WA *Statistical Register*

Teetulpa. As the new syndicates moved in, several previously unprofitable mines consistently recovered 80 per cent of their gold by cyanide; now only 20 per cent was coming from the battery.[12] Much closer to Adelaide, the government's own venture at Mount Torrens was intended to trigger another ripple of mining and employment.[13] From April 1894, its collective treatment facilities restored viability to surrounding mines. The government later repeated the pattern of filling private gaps with public cyanide works at Petersburg, nearly 150 miles north of Adelaide; at Arltunga, near Alice Springs; and at the isolated Tarcoola, 350 miles northwest from Adelaide.[14]

In New South Wales, the AGRC lacked the ready connections that had made Queensland and South Australia seem more hospitable. Perhaps this was why its own efforts there lacked a certain drive. With neither office nor custom plant to spark interest in a stagnant climate, the process was slow to gain a foothold. A plant erected in 1893 to treat the pyritic and copper-laden tailings at Mitchell's Creek induced no adjacent activity and was idle by late 1894.[15] Eventually the process became concentrated in

Figure 11.4   Cyanide works, Charters Towers, in the late 1890s (W. Lees, *The Goldfields of Queensland. Charters Towers Goldfield*, Brisbane, 1899, Mitchell Library)

the central west, especially around Parkes. The advance of Cobar was also notable after works established in 1896 at the AGRC's own mine generated a subsequent cluster of cyanide plants. Other centres emerged around Hillgrove to the north, and Pambula to the south.[16]

Indications are that most of the New South Wales plants were rather small operations: a few English companies, like Myall's United and its £12,000 plant at Dubbo,[17] made stark exceptions. Estimates based on cyanide imports suggest that together the New South Wales cyanide plants were producing about 14 per cent of the state's gold production in 1903, perhaps 20 per cent by the peak year of 1909.[18]

Victoria was also slow to warm to cyanide, despite early signs of interest and vast accumulations of tailings. Economic depression had combined with the poor performance at Daylesford and a Government Analyst who continued to deny its relevance. Plans made in 1893 by the Department of Mines to erect a plant for large-scale cyanide experiments had still not been acted upon when Newbery died in 1895.[19]

The year 1896 saw increasing attention to the process as the department suddenly became much more active in its promotion. A trial cyanide plant was installed, reports on the process were published, experiments were conducted. Under Newbery's open-minded and tenacious successor, more promising results were achieved, such as the Axedale tailings assaying at only 2 dwt 15 gr per ton which were eventually made to yield a profit of 6s 5d per ton out of a total extraction value of 7s 9d.[20] Suddenly the colony was alive to opportunity, as cyanide took off simultaneously at several different mining centres. At first the activity was confined to old tailing heaps, then increasingly applied as standard treatment for new tailings coming from the battery after amalgamation. Yet chlorination hung on: though almost totally displaced for new and old tailings, it retained its hold on the treatment of pyritic concentrates.[21]

A short-lived AGRC office in Melbourne seemed to have little influence on adoption.[22] Three jumps in the use of cyanide in Victoria—in 1896-97, in 1900-01, and in 1906-07—coincided with three specific actions of government to promote mining. The 1896 Mining Development Act provided for loans for the installation of plant for 'pioneer mining', and for testing the value of metalliferous material.[23] This was followed up by the active promotion of the Act and the benefits of cyanide by the Government Geologist on his lecture tour of 1897.[24] The effects were seen in the estimated £100,000 invested in cyanide plant by 1899, and the 805 employed in cyanide work.[25]

The second lift to cyanide production coincided with the government's purchase of the patent rights in 1900 and the reduction of the royalty to only 2.5 per cent. The third jump followed the complete elimination of the royalty at the end of 1905. The Department of Mines noted an immediate effect from this change, as low-grade material formerly considered too poor for handling shifted into the range of viability.[26] From this time the number of cyanide plants rose as their average yield steadily declined: the low-grade era had arrived.

Western Australia was destined to be the colony where cyanide would contribute most. Though it was a late entrant to the industry, spectacular finds at Coolgardie and Kalgoorlie in 1892–93 set it on the world mining map, and by 1894 it was 'all the rage' in London.[27] In May the London office of the AGRC appointed McBean Bowder and Co., a young offshoot of McIlwraith, McEacharn and Co., as the local agent. The colony's surging gold output soon took the Australian lead, and the use of cyanide ran in close parallel. First it was evident in the increasing number of cyanide plants, tacked onto one-quarter of the mines within ten years of cyanide's introduction to the colony. As mines went deeper, the process was also extended to a greater proportion of material.

The concentration of cyanide use on the so-called Eastern goldfields was notable, especially at East Coolgardie where Kalgoorlie, the Golden Mile, and eleven of the sixteen principal mines of the colony were located. In 1903 the cyanide output of the top ten producers of this field was contributing an estimated minimum of 25 per cent, possibly 30 per cent, of the peak gold production of Western Australia.[28]

This increasing use was strongly conditional on adaptation of the process to the field. The experiment involved did not come cheap. By the end of 1898, the combined capital of the operating companies at Kalgoorlie was £30 million,[29] and by 1902 over £8 million had been spent on metallurgical works, much of it 'devoted to the acquirement of experience';[30] some companies were by this time operating on their second or third plant. Since 163 of the 197 mining companies registered in 1897 were foreign,[31] the bulk of the finance for this experimentation came from overseas, as did much of the expertise.

Making it work, making it pay

By the end of the century the cyanide process was well established on Australian goldfields. But it had not been an easy run. A legal and political onslaught on the patentees had improved the terms. But something much less visible had also been taking place, as some tinkering here, some experiment there, gradually moulded the process to new landscapes. Now it seemed far more congenial to local ores, and more of them. Now its economic claims sounded far less hollow. Yet still, each colony had a different tale to tell.

Limits to innovation in Queensland

One early advance made by the AGRC in Queensland was in proving the viability and economic attraction of percolation over agitation. Even in 1891 this meant a cost of around 9s per ton as against 25s. The economics of adoption were also assisted by the 1896 reduction of the royalty to 5s per ounce, by the virtual halving of the price of cyanide, and by the accumulation of experience which brought more economy in the use of the cyanide salt itself. By 1900 the average cost of ordinary cyaniding at Charters Towers was quoted as 7s 6d per ton including royalty.[32]

The now standard practice involved several steps:[33] crushing, amalgamation on copper plates or mercury wells, concentration, grinding and amalgamation of the concentrates, concentration after grinding, then treatment of the whole or part of the residues by cyanide, usually by percolation. Incremental modifications were evident in the minor variations to percolation described by various users. On the Etheridge field, some bold experiment was evident in attempts to tame the copper-bearing ores by precipitating the copper before treatment.[34] But the complications of what had become a rather messy procedure were apparently left untouched.

The irony was that cyanide had offered such simplicity. Ample demonstration had shown that cyanide could be applied to the whole ore, direct from crushing by the battery. This would eliminate all the intermediary processing that had become typical of Queensland and, in particular, Charters Towers. Direct treatment would also allow for coarser, and therefore quicker and cheaper crushing. But the opportunity was ignored, despite restructuring of the cyanide royalty in a way which would encourage direct application of cyanide to the ore.[35] Why was this so?

Two factors seemed to conspire against vigorous experiment and extension of the process in Queensland, and particularly in Charters Towers. One was the uniform richness of the field: it consistently yielded handsome returns despite continuing lack of attention to inefficent crushing operations. The second was the social structure of the field, for this was dominated by the central crushing mills which serviced a mining community made up largely of individual miners or small self-financed syndicates. The mills had plant committed to grinding processes which they were not eager to abandon when the costs could simply be passed on. The miners had neither the volume of output to warrant a plant, nor the means or knowledge for experiment. Instead, they sold their tailings to the mills or to the operators of the small and crude cyanide plants set up to work sun-dried old tailings retrieved from a nearby stream. These were quite unsuited to direct treatment of freshly mined ore from a battery.[36]

So it was that cyanide was able to displace chlorination but not the series of pre-treatments which had become the habit of the field. It remained a process tagged onto the end. Operations at Charters Towers remained organised around the central mills, and cyanide treatment remained structured around practices which had been set in place twenty years before. Confined largely to treatment of accumulated tailings, much of which were discards from the past, the process declined as the tailing heaps levelled off.

Research, adaptation and diffusion in South Australia

Though gold and cyanide gave less to South Australia, that colony's capacity for experiment contributed to goldmining both within and without its borders. For one of this colony's assets was its Government Analyst, G. A. Goyder, the eldest son of the former Surveyor-General, renowned for his Goyder Line, which delineated the arid areas.

Goyder's education had been principally by private tuition, followed by wide practical experience.[37] Lessons learned in his own private assaying practice in Adelaide were broadened in the Moonta mines assay office. From there he was seconded by the government to assess some of the rabbit extermination methods thrown up by New South Wales' great contest of 1888. After a time spent supervising the local manufacture of carbon bisulphide, he was appointed Government Analyst in the Crown Lands Department and promptly produced the *Prospector's Pocket Book* to help the search for minerals. When his section was transferred to the new School of Mines which opened in Adelaide in 1889, Goyder was still only thirty-four years old, but he was about to gain the added duties of lecturing on assaying, chemistry and metallurgy. Soon he would show how an imported technology could propel local research through screening and adaptation and on to the mysteries of chemical mechanisms.

It was J. V. Parkes, Inspector of Mines, who started it all. He believed this new cyanide process had great potential, but he also knew that his predecessor, David Rosewarne, was bitterly disappointed at the high consumption of cyanide by his samples of Westward Ho ore. It seemed copper could be the culprit. As Parkes quizzed Goyder on copper's interference with the cyanide reaction, he did not know that he would spark a process of inquiry which would continue for several years. For Goyder could not find the answers he needed in the literature, at least in what was available

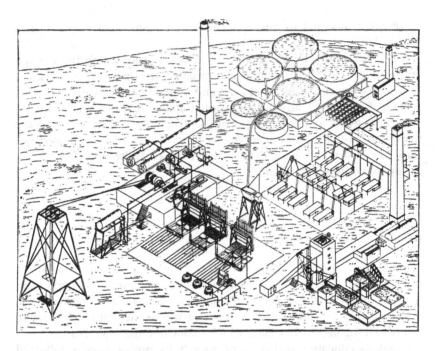

Figure 11.5   The cyanide plant at the South German mine, Maldon, Victoria
(*Australian Mining Standard*, 5 August 1897)

to him. There was no alternative but the lab and the test tube.[38]

His starting point was John Macarthur's statement that 'sulphides of copper, zinc and iron artificially prepared are readily acted on by cyanide of potassium, but not the mineral'.[39] But Goyder found that Macarthur was wrong: a range of natural mineral sulphides of copper were readily acted upon by cyanide. His conclusion was that potassium cyanide lost its power of dissolving gold when saturated by a copper mineral: clearly significant for a colony known for its copper-bearing ores. The conscientious inspector soon wanted tests on a wide range of local ores as he assessed the feasibility of government cyanide works. The positive results enabled Parkes to persuade the Commissioner of Crown Lands to proceed with the works, while disheartened miners were encouraged to guarantee their patronage.[40]

In the meantime, Goyder delved into Macarthur's claim that Elsner had been wrong back in 1846 when he said oxygen was necessary to dissolve gold by cyanide. Again Goyder found himself disagreeing with Macarthur. Experiments begun late in 1892 showed instead that gold and silver would dissolve in potassium cyanide only in the presence of oxygen, and that hydrogen was never evolved. This was no question of esoteric curiosity, the role of oxygen being acknowledged as of the greatest practical importance in the application of the process.[41] As such, it was to attract considerable attention in world-wide research, and Goyder established himself as a forerunner. His work was even cited in a German publication in 1896.[42]

During early 1894 Goyder was asked to assist in the remodelling of the Mount Torrens cyanide plant. The appointed manager, Lawrence William Grayson, had been in the first intake of students at the Adelaide School of Mines in 1889 and the first student to receive the Associate Diploma in 1891. Too early to have received cyanide tuition, he benefited from Goyder's experience. Determinations of cyanide solubility of several metals were found to have consequences for operations. By experimenting with the process, Goyder developed practical procedures covering the crushing of the ores, their exposure to the cyanide, and the optimum strength of cyanide solution under various conditions, all of which led to important alterations in plant design. Continuing trials on the method of leaching resulted in refined techniques for filling the vats with the tailings and cyanide solution, for running the liquid through the tailings, and for washing the cyanide out again. With improving extraction rates, even on formerly intractable ores, the government cut treatment charges from 12s 6d to 10s per ton.[43]

Improvements to process control came through apparently simple means, such as better titration to measure the amount of 'available' potassium cyanide in the presence of double cyanide salts. This in turn revealed that the acid needed to neutralise a solution before its passage to the zinc boxes was proportional to the amount of free cyanide, and provided a quick and effective index of the cyanide needed to leach the ore. Insights into chemical mechanisms came from finding that the solutions exiting the zinc boxes contained practically no simple cyanide of potassium and deducing that it was all converted into double salts.

To Goyder it seemed that the double cyanide salts formed during the course of the process, as cyanide acted on other metals in the ore, must also be able to dissolve gold. Back in his laboratory at the School of Mines, Goyder put this hypothesis to the test. He concluded that these double salts and their solvent action on gold could claim a large part of cyanide's success. Again he was at the forefront of world research. William Feldtmann, chemist to the South African Gold Recovery Co., was making a similar discovery that gold was soluble in a solution of the double cyanide of zinc and potassium. When reported in September 1894, it was regarded as the first statement of this 'paradoxical' fact.[44] By this time Goyder had already passed on results and recommendations through the Fowler family in Adelaide to the AGRC in London and to Macarthur and the Cassel Company in Glasgow.[45]

More experiments in 1895 continued the incremental refinements. First the injurious effect of iron in the zinc boxes was eradicated. Then the Minister's concern at the mint and transport charges on impure gold was soothed when a better way of purifying the final gold sludge pushed the value of bullion from £2 to £3 7s 9d.[46] Next it was found that air under pressure could hasten the rate of percolation and the rate of solution of the gold. As teething problems at Mount Torrens eased into smooth and successful operation it was reported that cyanide had made a 'great difference' to mining in South Australia.[47] With 23 per cent of the state's gold coming from several government works in the new century, government policy seemed vindicated.[48] Yet, perhaps even more significant than the direct boost to gold recovery was the demonstrated viability of cyanide and the indirect stimulus to mining. This was evident in the increasing flow of samples for gold assay at the School of Mines and the increasing buzz of private mining activity around the government works.

Goyder's work had other outlets through his teaching and affiliation with the new South Australian School of Mines. Its annual reports from 1893 carried detailed descriptions of experiments and results, and student knowledge of cyanide technology was tested by 1894 or earlier. The examiner was Edward Rennie, the chemistry professor at Adelaide University, a former research colleague of Goyder and a member of the school council. No doubt other councillors, including government representatives, were early eavesdroppers on Goyder's work.

To other scientific and technical colleagues, Goyder spoke through the meetings and journals of the Royal Society of South Australia and the Australasian Institute of Mining Engineers. Another School of Mines councillor and Adelaide University geologist, Professor Tate, introduced Goyder's presentation to the new institute in 1893 with the prediction that the process would have an important bearing upon the commercial aspect of mining.

Two years later, the institute's president acknowledged the bearing of another Goyder paper on the current uncertainty about how to work the process with success. The secretary saw value in the specific details on Mount Torrens and the revelations about the varying solvency power of cyanide on gold. Victoria's Thomas Davey looked upon the paper as 'one of the most important as affecting the mining industry, particularly the duties of a mining engineer'.[49] Arthur Holroyd was engineer at the Austral Otis Co. and now offered to test ore samples at a new cyanide test plant just erected there by the AGRC. W. B. Gray was especially attentive since his own experiments comparing cyaniding and chlorination in Victoria had led him to conclude that not enough was known about cyanide to use it on raw ores. He found Goyder's paper particularly valuable with regard to his difficulty in cyanide treatment of slimes. By 1896, Gray's cyanide plant at the South German mine at Maldon was fully operational and in 1897 Gray was himself presenting a paper to the institute on cyaniding slimes.[50]

Experiments, debate and economics in New South Wales

Between the first faltering moves toward cyanide at Mitchell's Creek and the burst of interest in the central west in 1896-97, local experiment was also taking effect in New South Wales. Australian consulting metallurgist, Charles A. Mulholland, Ph.D., got busy in his Bathurst laboratory in mid-1893. He wanted to adapt cyanide to local ores, but he needed to know how the chemistry worked. Macarthur's claim that oxygen was unnecessary had already sparked international research on the role of oxidation in gold solution, which was in turn a springboard for experiments with *other* oxidising agents which might supplement cyanide action.[51] Mulholland thought bromine or chlorine might speed cyanide's work.

He found that bromine could displace cyanogen from its compounds in the presence of oxygen and a metal. The result was a double cyanide of the metal and a bromate of the halogen. Hoping now to substitute bromine for oxygen in a cyanide process, he devised a method using free bromine in addition to potassium cyanide. A great variety of ores, tailings, concentrates and slimes were obtained for testing through 'many friends and mining managers' including F. M. Drake, of the Wentworth Gold Fields Proprietary Company. By early 1895, Mulholland claimed that results in most cases were good, in some cases very good, extracting 97 per cent in a time impossible with existing methods.[52]

Unfortunately, Mulholland's research was overtaken by the surprise announcement in 1895 of the recent patent by British workers H. Sulman and L. Teed, who had used *bromocyanide* to dissolve the gold. Finding that free oxygen went against 'previous experience and . . . scientific theory' and actually *prevented* the solution of gold by potassium cyanide, they aimed to substitute cyanogen bromide for oxygen.[53] The result, they said, accelerated gold solution from hours to minutes. To Mulholland it seemed the best course was to release the results of his own incomplete research 'untrammeled by patent rights'. Through the *Australian Mining Standard* he claimed his method used less cyanide than the conventional cyanide process, while giving more rapid solution and a higher rate of recovery. It would help those whose ores, tailings, or slimes had hitherto resisted economical extraction.

Close on his heels, Dr John Storer revealed similar experiments to readers of the *Australian Mining Standard*. As an immigrant scientist with Scottish and German training in chemistry and mining, and abundant international consulting experience, Storer felt he had much to offer the local mining industry. He had already patented the 'Storer-Marsh' treatment of Broken Hill sulphide ores, and since 1887 had been trying different approaches to gold and silver extraction.[54] Now, he too, had come to the use of bromine, but thought his way was better. Against the Sulman-Teed method, he argued that his use of bromine was cheaper and more safely handled than their use of cyanogen bromide. Against Mulholland, he argued that direct addition of bromine to the ore was wasteful, a problem avoided by mixing the bromine and potassium cyanide *before* adding to the ore.[55]

The debate between Storer and Mulholland intensified when they broached the question of how the reaction worked. Storer had the simpler view and argued that no oxygen was required. Mulholland argued that *in practice*, parallel oxidation could not be eliminated so that, in reality, the extraction of gold must be attributed to a combination of bromination and oxidation, even though the aim was to promote the former over the latter. Their differences were exposed and debated in a series of correspondence which detailed practice, experiment, equations and theoretical positions.[56]

The readers of the *Australian Mining Standard* were impressed. Many wrote for further details. The extent to which the recommendations were put into general practice on the goldfields is difficult to discern, but by late 1896 Storer was certainly involved in bromocyanide experiments at the South Star mine at Ballarat, where approximately 30 per cent of the gold was being carried away in the tailings in the form of fine gold, causing losses worth £24,000. Storer was then engaged as consulting chemist and engineer to direct the works when the company adopted the bromocyanide process on agreement with the Sulman-Teed patent holders.[57]

Back in New South Wales, Mulholland's earlier experiments with the Wentworth Gold Fields Proprietary Company at Lucknow may have been responsible for that company's routine use of the cyanide process in some form by at least 1897. Not that their problems were all solved. Continuing difficulties with gold-laden slimes led to continuing experiment and plant modification. By 1899, the cyanide manager, Henry Watson, could claim 90 per cent extraction rates and predict that his 200,000 tons of tailings, averaging 8 dwt per ton, would yield 80,000 bullion ounces at £3 10s per ounce—a total of £280,000.[58]

Mulholland's work was also propagated through other channels. Locally, his election in 1897 as president of the Bathurst Science Society reinvigorated the society's meetings at the Bathurst Technical College. With scientific contributions such as his 'Discourse on a method of treatment for low-grade auriferous copper ores', the society regularly attracted audiences of fifty to sixty. It was such work which led to his election in 1901 as chairman of the new Sydney Technical College Mining and Metallurgical Society. His first paper to it was, appropriately, 'Some theoretical considerations in cyanide practices', a topic considered worthy of full publication in the *Australian Mining Standard*.[59] Internationally, Mulholland's paper to the prestigious *Engineering and Mining Journal*[60] injected his findings into the international arena, where they were taken up by other mining scientists as part of the important debate on the mechanism by which gold was dissolved by cyanide.[61]

It might have been the debate in the pages of the national mining journal during 1895 which stimulated Goyder in South Australia to pursue some of his own investigations into the bromocyanide reaction. He could see it impinged on his own observations regarding the accelerating effect of the double cyanides. He was led to the conclusion that two steps were involved in the process, and that the enhanced solubility observed with bromocyanide was due to the gold existing in both aurous and auric states and readily passing from one state to the other under favourable conditions. Perhaps Inspector Parkes, who had initiated Goyder's cyanide research, followed these further developments of his work. If so, it would have proved invaluable when he moved to the west to become manager of the Oroya Brownhill mine at Kalgoorlie. Oroya became a large user of the cyanide process and eventually the bromocyanide modification.

Incremental improvements and low-grade ores

While all this investigation of mechanisms had been taking place in laboratories, however, another process had also been at work at the plants—a steady chipping away at the economics of the process by small, incremental changes. The working of the Mitchell's Creek tailings is typical. Here a few years would make a difference to the economics of operations handicapped by the presence of copper. By 1904 there were 200,000 tons of tailings in the form of sands and slimes, which it was said had so far resisted successful treatment on a large scale because of the iron and copper pyrites. Their gold content was now on average lower than in 1894, tailings coming from the battery usually carrying just over 3 dwt of gold per ton, valued at 13s. By this time, however, the cost of treatment had also come down. Wilfred MacDonald, a mining student, had carried out numerous experiments, finding that double leaching could give a 55 per cent extraction on an experimental scale, with 65 per cent anticipated on a commercial scale. Though this extraction rate was similar to that achieved in 1894, and again in 1898, working expenses were now expected to be 5s per ton, leaving a clear profit of 3s 6d. On this basis, in May 1904, the mine manager recommended erection of a double-treatment plant capable of treating 1000 tons per week.[62]

Much of the cyanide processing was on low-grade ores. The cyanide plant of the Myall's United Gold Mine, near Dubbo, was in 1899 working profitably on 6 dwt ore, 4 dwt coming from amalgamation and 1.5 dwt of fine gold by cyanide at a total cost of 13s 10d per ton. Cyaniding accounted for only 3s 2d per ton.[63] At Yalwal in the south and Hillgrove in the north, tailings of 5 dwt were yielding profitable extractions, while out near Parkes recoveries of 2.5 dwt per ton by *direct* cyanidation were found worth the expenditure of £2000 on cyanide plant.[64] If the process was contributing something like 20 per cent of the gold output when cyanide imports peaked in 1909,[65] it seems that this had largely come about through the steady erosion of costs by experiments both in private laboratories and in user plants. High-grade ores could also benefit, but the real significance of cyanide now seemed to lie in tipping the balance of profitability on ores previously regarded as too poor or too troublesome to treat.

A centre of innovation in Victoria

In Victoria, where fine pyrites gold escaped amalgamation, and also evaded Duncan McIntyre's early percolation plant, a cluster of experimental activities attempted to deal with the slimes by agitation. It was the fineness of the crushed ore coming from the battery which hindered percolation: the slimes settled so densely in the vats that the liquor could not seep through. The percolation problem could be solved by first running the battery tailings through a Butter's distributor to waylay the very fine slimes, but this lost a portion which often held a high gold content.[66] It was therefore significant when users devised their own innovations to resolve these physical problems.

A Scottish-educated engineer, George Duncan, had been chief engineer to the Melbourne Tramway and Omnibus Co. before going into private practice. He was using cyanide to treat tailings from 1896 and designed machinery which was then manufactured by his firm of Duncan Noyes and Co., who also purchased tailings to treat by cyanide under a special arrangement with the AGRC.[67] But it was John J. Deeble of Bendigo whom the *Australian Mining Standard* dubbed 'the pioneer of cyaniding in Victoria'.[68] He had been been treating pyrites by various processes since 1871, and in early 1896 turned to cyanide. In the pursuit of maximum extraction, he devised his own machine to deal with the

slimes. The 'Deeble Patent Agitator' consisted of a circular vat with agitating blades capable of being raised or lowered to keep the whole mix
thoroughly stirred. Afterwards, the slimes were allowed to settle and a
side gate gradually lowered to allow the clear liquor carrying the dissolved gold to flow out.[69]

Over a period of eighteen months, Deeble profitably recovered just
over 6 dwt per ton from 20,000 tons of abandoned tailings. Soon he
moved on to 4-5 dwt pyritic slimes previously thought unpayable.
When he applied his techniques to an unroasted sulphide ore, he recovered 87 per cent of its gold. With even finer grinding he increased this
to 95 per cent, leading to a claim that any refractory ore ground to a
pulp and treated by his machine would yield 90 per cent of *all* its
pyritic gold without prior roasting. When the judges of Melbourne's
1897 Mining Exhibition put this to the test on refractory ore assaying
at 7 oz 2 dwt of gold, the assay of residues after the 23-hour treatment
indicated an average recovery of 86.5 per cent. Much impressed, the
judges awarded Deeble a First Order of Merit and directed the attention
of the mining community to the virtues of the innovation. By the end
of 1897 there were reportedly nine of these plants in the Australian
colonies, including two in Western Australia, and at least one in New
South Wales at Wyalong.[70] A visiting British mining expert observed
late in 1897 that the Deeble system had recently come into 'considerable use'.[71]

One of the first adopters was W. B. Gray, who had been experimenting with slimes since 1895. His South German mine at Maldon
was a substantial enterprise, rating third among Victorian mines both
as a gold producer and dividend payer. Gray roasted and chlorinated
his concentrates, but by 1896 was cyaniding his tailings. Having to
resort to the Butter's distributor to separate the troublesome slimes,
Gray had been recovering an average 2 dwt 7 gr per ton of tailings,
but installation of the Deeble system in mid-1897 allowed him to
recover 92.5 per cent of the 9 dwt of gold in each ton of the previously
discarded slimes. In anticipation of even higher future results, Gray
installed four extra cyanide vats.[72]

A complementary innovation came from Deeble's experience with
the Newbery–Vautin method of chlorination, which used charcoal filters
to precipitate the gold from solution. In Britain late in 1897 the novelty
of using this in combination with the cyanide process was noted.[73]
Both aspects were made available to the Ballarat School of Mines when
in 1898 Deeble donated his complete Model Agitation Plant to com-
plement their existing percolation plant.

The other innovation to come from Victoria at this time was a joint effort
by David White and Thomas Moore Simpson of Stawell. David White was

an engineer with long experience, having for some years represented the Luhrig Ore Dressing Co. of London. Simpson was a local metallurgist, trained at the local school of mines. In 1896 they set up an experimental plant at Stawell to process slime and fine pulp. Conical vats held perforated bottoms through which compressed air was blown to maintain thorough agitation and remove any film from the gold, exposing surfaces for the cyanide to attack. To prevent loss of cyanide from the passage of air, the vats were arranged in a series, so that the air from the first vat passed through the second, and so on to the end of the series, and then back to the first so that it was continually reused. Once the air became saturated with cyanide vapour, it ceased to cause any further loss through decomposition.[74]

In their patent specification, White and Simpson claimed to have solved the slime problem which had for many years engaged metallurgists and chemists, and this in hours rather than the days required by percolation. This speed, they claimed, was also responsible for the marked economy of cyanide consumption which accompanied their high extraction rates on a wide range of materials. While no statistics indicate the use of the White–Simpson system, the Victorian Department of Mines installed a small plant in 1897 for the experimental treatment of slimes. This complemented their percolation plant, and both were used to advise upon the suitability of cyanide treatment for a large number of samples of Victorian ores. By the end of 1898 the Secretary for Mines was commenting upon the remarkable and sudden change of attitude and activity directed to maximising the extraction of gold in Victoria. The shift was clearly due to the proof of what could be done with even very low-grade material. As evidence that high-grade ore was not indispensable to profitable working, the secretary cited Stewart's United Co. at Bendigo which had been able to pay a dividend on a yield of only 2 dwt to the ton. By 1900 some companies were able to pay a dividend on a yield of only 1.5 dwt. By 1901 others were reporting profitability on a gold content of only 10 grains.[75]

The 'Westralian' pattern: idiosyncrasy and innovation

Western Australia had struck it rich with the gold finds of 1892–93, but these alluvial deposits were shallow, exhausted within two years. The future now lay in the lodes. On Norseman, Gwalia, Coolgardie and Big Bell, cyanide could routinely flush out the 25 per cent of gold left in the tailings after amalgamation,[76] but Kalgoorlie ores posed a greater challenge. Even in the oxidised layers, gold particles were so naturally fine that metallurgists constantly struggled with slimes. Below 200 feet lay the puzzling sulphide ores which were predominant by the early years of the new century. Then there was the rare and valuable telluride of gold, first revealed in 1896. High lime and magnesium levels made its chlorination impracticable, yet normal cyanide solutions proved poor solvents.[77]

Location only added to the difficulties: the total absence of fresh natural water, the limited supply of fuel, the high freight costs, high import duties, high cost of living, and consequently high wages all contributed to the Kalgoorlie imperative to innovate. In this, the mines could learn from each other, but imitation was limited by the lack of uniformity in the mineral contents of even the closest deposits.

John Macarthur and Cassel colleague, John Yates, saw first that the cyanide process could be adjusted to suit this exciting new goldfield. Their Barrel Amalgamating Apparatus rolled cyaniding and amalgamation into a single process, minimising water and fuel requirements, as well as handling and labour.[78] Though it impressed London experts, its failure *in situ* at Hannan's Brownhill in 1896 caused a general setback for cyanide at Kalgoorlie.[79] But only a year later, Lake View Consols gave cyaniding a second chance after finding that 50 per cent of its gold was now escaping in the tailings. Many others were soon in the same position. By mid-1898 cyanide was reportedly extracting one-quarter of Kalgoorlie's gold output and a period of great experimentation had begun.[80]

Adapting to the slimes
The first hurdle was the slimes, a product not of crushing but of the naturally high proportion of fines in the ore. Averages of 50 per cent could block the percolation of cyanide even through coarsely ground material.[81] The South Africans had found decantation a useful method, but it was unsuited to Kalgoorlie's rich ore and lack of water: experiments showed the settling of particles inconsistent and unreliable.

The Westralian way was found by Lake View Consols when their Australian metallurgist, John Sutherland, toyed with the filter press used in sugar mills to separate juice from pulp. The idea was not new, having been tried in the Cassel works in Glasgow and New Zealand. But there was significant innovation in the way Sutherland pumped slimes direct into the presses, forced the cyanide solutions through the cloth to dissolve the gold, and then forced water through to displace the gold solution.[82] In early 1898 he increased the extraction from tailings by 50 per cent, taking cyanide contribution to 52 per cent of total gold recovery.[83]

Early treatment costs worked out at 10s 4d per ton, but by 1900 the Lake View plant could process 4000 tons of slime a month for 6s per ton.[84] In the interim, other companies had experimented with specific designs and operations, but broadly, two methods had evolved. Single pressing involved prior solution of the gold in cyanide, using the press only as a filtering medium, or alternatively, bringing the gold into solution in the

Figure 11.6   Workers with a filter press at Lake View Consols (A. G. Charleton, *Gold Mining and Milling in Western Australia*, London, 1903, Mitchell Library)

presses themselves. In double pressing the slimes were forced through the presses both before and after cyanide treatment. Although slightly more expensive, it was most likely to give the highest extraction, but the choice depended on the nature of the ore and the method of crushing.

The filter press won favour because it allowed cyanide to be applied to the full range of tailings coming from the mill, dramatically reducing the loss of fine gold. The fact that it did so in an operation that was completed in only two hours attracted almost universal adoption on the field. Donald Clark reflected on the consequent irony that 'the terrible bugbear of the old-time amalgamators and concentrators should be looked upon as a blessing by modern metallurgists. Avoid making slimes was the old, and slime everything the newest doctrine'.[85]

Adapting to sulphides and tellurides
Still the chemical problems remained. Normal sulphide ores were difficult enough but sulphotellurides were worse, tests showing that only 20 per cent of the gold was recoverable by ordinary methods.[86] One factor was the variability of telluride concentrations, from zero to 3 per cent.

From 1897, experienced metallurgists at all the principal mines were engaged in experimental work on 'the sulphide problem'.[87] Straight cyanide was a poor solvent for the sulphotelluride; chlorination was ruled out by the high levels of lime, magnesia and alumina; and shipment for smelting in South Australia was extremely expensive, typically more than £5 per ton. Gradually, the idea emerged that light roasting would break down the tellurides, converting their gold to the metallic state and their solid sulphide particles to porous oxides, enabling subsequent cyanide treatment to penetrate and attack the formerly enclosed gold.[88]

This method took the name of its pioneer, J. T. Marriner, the Cassel-trained metallurgist at the Boulder Main Reef. It derived from research carried out at the Cassel works in Glasgow, but in practice several difficulties were encountered. One was due to the poor working of furnaces unsuited to the peculiarities of the ore. The Boulder worked its way from the 'Mount Morgan' multiple hearth furnace to a design from Edwards of Ballarat, then another from Mertons' metallurgical works at Spottiswoode. At Lake View Consols, mechanical problems with the roasters in the latter part of 1899 were approached by engaging several experts (also unsuccessful) from Swansea in Britain.[89]

Roasting also had the disadvantage of forming much coarse gold, which then required lengthy exposure to cyanide. An aggravating factor was the setting that was caused by the partial conversion of the lime content to calcium sulphate. It seemed simple enough in retrospect to find that pulping the roasted ore, then amalgamating in a grinding pan, would inhibit setting, enhance grinding capacity and amalgamate the coarse gold. If the pulp from the grinding pans then went through the filter presses, it was possible to slash the high cyanide consumption caused by the soluble sulphates formed during roasting.[90]

The Dhiel process

Some metallurgists preferred to work on the chemistry of cyanide. At the London and Hamburg Gold Recovery Co. at Kalgoorlie, Dr Ludwig Dhiel and H. Knutsen conducted trials on a range of local sulphotellurides and revealed that extraction over twenty-four hours could be almost doubled by the addition of bromocyanide. The resulting 'Dhiel Process' extended the use of bromocyanide to the previously insoluble telluride of gold.[91] It included three essential stages: crushing and sliming the ore; treating the slimes in agitators with a solution of potassium cyanide in conjunction with cyanogen bromide; filter-pressing the sludge, and precipitating the gold from solution by means of zinc shavings. Sometimes amalgamation and concentration were included. The result was a process which effectively applied cyanide to the whole of the ore.

In 1902 eleven Kalgoorlie mines were treating sulphotelluride ores and three were using Dhiel's method.[92] At Lake View Consols, frustrations with the failure of Swansea furnace experts led first to a small trial, then a decision to run bromocyanide routinely in parallel with the existing roasting plant. Hannan's Brownhill followed soon after, with a completely new plant capable of treating 50 tons of sulphotelluride ore per day. Hannan's Star had modified its milling plant several times and was recovering 92 per cent at 21s 4d per ton. By 1905 the new metallurgist, E. W. Nardin, was proving that recoveries increased in inverse proportion to reduction in alkalinity, that iron sulphide and metallic iron destroyed bromocyanide, and that lime could cause some loss of gold through reprecipitation. He was able to change practice to cut bromocyaniding costs to 9.75d per ton.[93]

As observed by the Cassel chemist, Alfred James,[94] the process eventually adopted at Kalgoorlie mines was an example of the experimental resolution of difficulties as they occurred, a case of learning by doing and of innovating in small increments. It meant most plants had their own version of the cyanide process, but it also meant costs were being generally reduced. The experienced Algernon C. Moreing, former president of the Institute of Mining and Metallurgy in London, could quote a whole list of mines where total costs fell by about 40 per cent between 1901 and 1904. Still the experimental work continued, and on 'scientific principles—not the usual plunging in the dark in the hope that the unexpected may happen'.[95]

An important element of this approach had been the extended application of cyanide's powers. In the 'continuous-process' plants operating at all the major Kalgoorlie mines in 1907, the ore was in contact with cyanide solution throughout the course of treatment.[96] It was one of the factors in the fall in overall operating costs of Western Australian mines from 58s 4d per ton in 1903 to 32s 5d in 1905, a reduction of nearly 42 per cent on the mining and treatment of 8 million tons of ore.[97]

The technical adaptations made in Australia's gold colonies were often mundane and individually unremarkable, but their accumulation gradually changed the cost structure facing potential adopters. The result was more than a greater payout to shareholders. It meant that large low-grade deposits could be mined, thus broadening the basis of the industry, enlarging the field for employment, and providing the benefits of greater production of wealth to the wider community.

Looking back from the first decade of the twentieth century, it could be seen that both the cyanide process and the mining industry had come a long way in Australia over the previous twenty years. Both had a long way still to go. Though a mining downturn was just around the corner, even bigger changes lay further ahead. For cyanide was not a passing process. It would continue to grow as one of the major means of gold recovery, especially for low-grade ores, as it moved through further generations of adaptations, including the carbon-in-pulp (CIP) cyanidation, which forms an important part of modern goldmining operations and contributed to much of the remarkable expansion of Australian gold production in the 1980s. The ability of the Australian industry to progress with the process depended, however, upon other, institutional, changes.

CHAPTER TWELVE

Out of the hands of 'rule-of-thumb men'

Out of the early failures to capture the deep reserves of Australian gold, an awareness gradually dawned in the late 1880s that there was 'something amiss'. The industry was abandoning mines simply because there was no method or qualified men for treating the ores.[1] As the cyanide process penetrated the mining psyche in the 1890s, it promised to solve the treatment problem; but it carried its own implicit demands for technological expertise and facilities. It was a message which slowly got through, driven in by the increasing evidence of what could be achieved by the application of scientific methods. A new century would see the cyanide process established as part of the scientific armoury of the mining industry and a more professionalised model of mining education as part of its technical infrastructure. An imported technology had left its mark.

Seeing the gap

In the wake of a now acknowledged gap in local capability, two kinds of solutions were aired in the late 1880s: one envisaged the foundation of government metallurgical works to fill the void of industry expertise; the other advocated mining colleges which could equip industry with the skills needed to tackle technical challenges itself. In some minds, the need for the works and the colleges were inextricably linked.

Such proposals were, of course, a 'departure from the ordinary and legitimate undertakings of the Government'.[2] But plenty of justification could be found, from the costly necessity to send refractory ores to America or Germany, to the protection of the investing public and the employment of the working miner. It was also alleged that private enterprise efforts in this direction had failed.

The specifics of proposals varied from colony to colony. New South Wales leaned toward metallurgical works staffed by imported experts.[3] Victoria and South Australia leaned toward a central college, modelled on the Royal School of Mines in London.[4] Queensland's parliament was less ambitious in its 1885 resolutions affirming that mineralogical lecturers should tour the country to instruct miners on the detection of valuable minerals.[5] This was much more in line with past traditions of training than those of the future.

Traditional mining education

There was a substantial problem hanging over the traditional mining education in Australia: it was largely perceived as a function of those culturally-inspired institutions which laboured to diffuse useful knowledge and aid adult education.[6] It took place in the lectures and classes of schools of arts, mechanics' institutes and the like, organisations usually initiated, organised and controlled by local committees. Even those bearing the title of school of mines exhibited little difference either in subject range, organisation or standards, from those of lesser pretensions. Funding typically came via local subscriptions, supplemented by formula from government, usually in the vicinity of 30-40 per cent. Courses depended on local predilections and access to teachers. Generally, with neither entrance or progression prerequisites, nor attendance requirements, the training provided was *ad hoc*, uncoordinated and unsystematic.

The School of Mines at Ballarat presented something of an exception, less so its counterpart at Sandhurst. But even though the programs there were more specifically and systematically aimed at miners, there was still much that kept them in the mould of the informal institutions of the cultural infrastructure. The Ballarat school had grown out of the Ballarat Mechanics' Institute, through the agency of the Ballarat Mining Board, Sandhurst out of the combined initiative of the local borough council and the committee of the local mechanics' institute. In 1888, of 799 students on the Sandhurst roll, only 10 per cent were miners, while only half were studying mining and scientific subjects. And the teaching staff at Ballarat and Sandhurst were well aware that students had neither the appropriate prior education, nor the length of attendance to acquire a proper technical education.[7]

Even so, they were ahead of the more typical servicing of miners' technical needs represented by the attempt to establish some technical instruction in Queensland, where A. W. Clarke took up his tour of duty as Mineralogical Lecturer from January 1886. Clarke's educational program operated at two levels—through popular overview lectures accompanied by some demonstrated experiments, and through 'hands-on' courses of three to five months. Miners usually predominated, but clerks, schoolmasters, tradesmen, storekeepers and officials were well represented. Students might attend about 50 per cent of classes, dropping off 'principally due to want of patience'. Accommodation was provided by perhaps negotiating the use of a shed with functional benches and shelves. Clarke would set up his own blowpipe table and appliances and a good assaying furnace. A Fletcher's benzoline blast furnace for smelting catered for advanced students who had already attended the blowpipe and wet testing courses. The suggested remedy for the lack of a good portable set of minerals for demonstration purposes was an appeal for local schools of arts to start mineral collections to foster an interest in mineralogy.[8]

As Clarke roved through Queensland's mining towns during the next two years, his path inevitably took him to Charters Towers where 2000 of the colony's 9305 miners were working at that time.[9] In 1888, one hundred people attended the blowpipe course, sixty-eight of them miners and another six proprietors of works and managers of mines. The miners averaged an attendance of 57 per cent of the fifteen lectures, while mine managers and owners averaged 52 per cent. The second course, on wet testing and the chemistry of minerals, was much less popular, allegedly because the necessary instruments were bulky, expensive and fragile; unlike those for the blowpipe course, they could not be cobbled together from items available in the village. Though thirty people attended the preliminary meeting, only eighteen turned up for the first lecture, including nine miners and two mine-owners or managers. But they at least saw out the course, with attendances averaging between 88 and 96 per cent.[10]

This 1888 visit prompted longer-term visions when a public meeting was held to discuss the idea of establishing a school of mines, based on an extension of the local school of arts. When the Government Geologist, Robert Jack, met with the organising committee, he interpreted their desire to start on a 'somewhat limited footing' as an initial requirement for instruction in mineralogy, chemistry, assaying, geology, mining and engineering, and for the construction of classrooms, laboratory and a museum. To this he added his own prescription that the school should 'not omit instruction in natural history and paleontology'.[11] As a Scottish-educated geologist of some scientific standing,[12] Jack took the whole project very seriously.

The model he proposed combined a systematic program of professional training with an experimental metallurgical works to service the needs of the mining community. He envisaged a school catering for two different levels of student. Those wanting a professional training would sit a prior matriculation examination to ensure an adequate starting level of competence, then complete systematic courses of instruction and examination to qualify as Associates of the School. Parallel provision

Figure 12.1   Robert L. Jack, Government Geologist, Queensland (*Australian Mining Standard*, 26 January 1899)

would be made for students who simply attended lectures as and when they desired. All students could receive, after examination, certificates of proficiency in any individual subject or of competency for certain positions, such as underground manager, engineer or assayer.

Staffing options revealed the limitations imposed by an institution conceived within the framework of the cultural infrastructure. For the present lecturer to teach mineralogy, chemistry and assaying he would have to withdraw from itinerant work. Mining and engineering could be taught by the Inspector of Mines. Only the work of the Geological Survey seemed too important for lecturing diversions, compelling appointment of special lecturers for geology, natural history and paleontology. More ambitiously, Jack considered a first-class Professor of Metallurgy, warranting a salary of about £1000, indispensable to the metallurgical works.

The funding formula for covering the school's annual running costs of £2700 was more typical—£500 might be expected from student fees, one-third of the remainder from public subscription, two-thirds from an annual government grant provided on the basis of £2 for every £1 subscribed. If this level of public support did not materialise, it could be concluded that the institution was not a public necessity. The proof came when the sum placed conditionally on the Estimates found no match in public donations.[13]

The next year a new itinerant Mineralogical Lecturer visited Charters Towers and faced the typical dilemma of how to instil theoretical principles as the basis of more practical techniques. Many students ceased to attend through their lack of appreciation of the inherent plan of the lectures, and their insurmountable lack of prior education. The only way William Thompson could see of catering for more advanced needs was to encourage a small core of active students to establish an institute to collect and discuss facts relating to goldmining.[14]

Efforts elsewhere ran a similar course. In Townsville, Thompson met the School of Arts' desire for technical tuition by arranging classes on geology and mineralogy. From one hundred people at a public meeting only eight enrolled and only four completed the course, the impression being that geology was a 'dry and difficult science'. At Gympie in 1890 a nucleus under the name of School of Mines was formed within the School of Arts, but in 1895 chemistry was the only subject related to mining. The year 1896 saw only three students in the mining surveying class, and by 1897 there were none. The mineralogy class was also devoid of students. There were fifty-one technical students, but the advertised technical classes included needlework, painting, drawing, French, singing and piano: only mathematics and surveying had any mining relevance.[15]

The rash of schools of mines emerging in Victoria in the late 1880s and early 1890s were little better. Queensland's Under-Secretary of Mines described them as 'a farce', while Mount Morgan's George Richard, a graduate of Ballarat, regarded them as positively dangerous.[16] In typical cultural infrastructural form, they fell far short of anything that could be taken seriously by those who were starting to articulate a systematic and scientifically based training for the production of mining professionals.

Filling the gap with government works

Into the breach stepped those governments concerned by the low technological capability of mining industry. In varying degrees, they were prepared to assume the cost and responsibility of delivering technical services to private enterprise, to encourage the working of deposits which might otherwise be left untapped. As the 1890s showed the cyanide process to be a significant part of the desirable armoury, the expertise and facilities needed for its use were identified as part of the gap government should attempt to fill.

It was this kind of thinking which set the South Australian government on the path to its own cyanide works at Mount Torrens in 1893, as already described. Designed to test and treat ores from surrounding districts, it was part of a package of initiatives intended to boost mining and the economy. Once successful, the strategy was repeated in the four government cyanide custom works spaced out across the colony by 1901 to serve small-scale mines in different gold-bearing districts.

The New South Wales government was less systematic and more tardy in implementing a policy articulated in 1889: by 1893 it had finally obtained an expert but no metallurgical works. An imported metallurgist, James Taylor, conducted assays around the colony and furnished reports on the treatment of refractory ores while impatient members of parliament awaited the opening of the Clyde works.[17] The cyanide plant finally installed there in 1897 treated 72 samples of 5–10 tons during 1898, while the associated chemistry laboratory experimented with cyanide treatment of 134 smaller samples.[18]

In Victoria, Cosmo Newbery's successor moved progressively into the investigation of the cyanide process and its applicability to Victorian ores. By 1897 the Department of Mines could offer the mining public the treatment of one-ton samples of tailings by percolation, or the agitation of slimes by the Deeble system. Business was brisk for a couple of years, then dropped off as more cyanide plants were added to the mining districts of the colony.[19] In several schools of mines, government-funded cyanide plants were filling a dual role of practical instruction for students and experimental treatment for miners. Their use by the public was evident in the 1898 complaint to Parliament about their inadequate scale, and the 1901 report of the Fink Technical Education Commission, which criticised the subordination of educational functions to commercial work.[20]

In Western Australia, the hearings of the Royal Commission on Mining of 1897–98 established general industry support for the government provision of public batteries, but more reticence about government metallurgical works for the treatment of refractory ore. One enthusiast was J. V. Parkes, former South Australian Inspector of Mines, now manager of Kalgoorlie's Oroya mine. He spoke from experience in recommending a government cyanide plant to demonstrate the best manner of telluride treatment. But the majority of witnesses agreed with Lake View's Henry

Callahan, who was adamant that where treatment works were needed private enterprise would provide them.[21]

Within two years perceptions had changed. As observed by mining expert Donald Clark, it was now well recognised that tailings from rich ore should never be allowed to run to waste.[22] By September 1900, cyanide works had been completed at the Norseman public battery at a total cost of £1232 19s 1d.[23] By 1903, there were eight public batteries equipped with cyanide plants, yielding a total output of 64,887 ounces. Of that, 11.3 per cent came from cyaniding and added £26,328 to its value.[24]

Government cyanide plants were intended to fill a gap in private extraction facilities at unattractive locations, but they also filled an accompanying expertise gap. In 1904, mining expert A. G. Charleton attributed the increased number of productive mines in several key districts to the facilities offered by government and private 'custom plants'.[25] Donald Clark, as a scientific man, believed these works should also be leading, not following, in methods of ore treatment, particularly for more complex and refractory material.[26] Certainly the talent for innovation was there, for the patentees of the White–Simpson method of cyaniding slimes were now highly placed in the running of the public system of cyanide works.[27]

In Queensland, Robert Jack's vision of a research-oriented metallurgical works seemed to pass into oblivion.[28] Yet the 1897 royal commission into the mining industry heard of many difficulties in the way of matching treatment to the variety of refractory ores. Suggestions as to how the government should fill this void varied from the provision of a qualified assayer and sampler, to the construction of a centralised metallurgical works, but most saw a cyanide works as at least part of the solution.[29] Though a commissioner implied that miners could combine to put up their own cyanide plant, one witness replied that it required expert labour, and the miners could not manage it alone. Captain Richard of Mount Morgan was even more emphatic about the extent and depth of technical assistance necessary for metallurgical works to be effective in practice. This was not yet the Queensland way, but the commission did recommend that the government provide a financial subsidy to enable miners to send difficult ore to private works such as those at Aldershot. There the technical expert from Germany provided free treatment advice to supplement the prime business of purchasing different classes of ores for treatment by the most suitable method, including cyaniding.[30] For the time being the Queensland government opted to pay for the cost of sending stone in lots of 5 or 10 tons to Aldershot from any part of the colony.[31]

Advancing cyaniding and capability

While governments toyed with ways to compensate for mining deficiencies, the industry's skills, resources and infrastructure were being nudged forward along with the perceptions and priorities of mining people. For with the use of cyanide a new technology was taking root.

From 1897 it was visible in the capital committed to physical plant on the goldfields, and then in the engineering firms that catered for that demand. Edward and Wallace Jacques were advertising cyanide plants among the engineering products of their Victorian Iron Works by 1896, and by 1898 had their own laboratory and testing plant where they assayed and treated ore samples by cyanide, chorination or battery.[32] The large Austral Otis Engineering Co. was performing a similar function from 1895, and in 1897 made a duplex compound steam pumping engine for Ballarat's South Star Company to go with the new bromocyanide plant.[33] While George Duncan was using cyanide to recover gold, his engineering firm, Duncan Noyes and Co., was making money from the design and manufacture of cyanide plants. That was why George Duncan took such an active role in fighting the AGRC's cyanide patents. In Bendigo, Henry Dunn and his partner had extensive experience in manufacturing mining machinery and by 1904 had executed large orders for cyanide plants for both Victoria and Western Australia, enough to keep twenty people constantly employed and to warrant substantial building alterations. An additional 'specialty' was the charcoal crushing and grinding apparatus used to provide charcoal for cyanide filter boxes. As the sole makers and patentees of this apparatus, the firm received 'a continuous demand from all the States'.[34] Though Australian firms were displaced from Western Australian fields when engineers from the Rand and America ordered heavier types of crushing machines from England and America, this prompted an 'awakening' of some of the Victorian and South Australian firms. In 1904 Donald Clark reported with pleasure that many of the latest designs had been adopted.[35] James Martin and Co. was certainly supplying all classes of mining machinery, including cyanide plants, from its Phoenix foundry at Gawler in South Australia.[36]

Along with the hardware of technology, a psychological commitment to new habits of thought, organisation and action was also infiltrating the goldfields. It came from two sources, the process itself and a few alert scientists. The most direct source was the process, pressing home its own messages to the industry as it exerted its own demands on users: only preliminary tests could determine the optimum method and conditions; routine laboratory process control could mean the difference between success and failure in the treatment of a particular ore; only adaptive experiments could manipulate the attack of active cyanide molecules on different classes and grades of ore.[37] The corollary was that the cyanide process carried an implicit prerequisite for specific kinds of expertise and facilities. In 1905, John Macarthur reflected that where 'rule of thumb had reigned supreme' cyanidation had compelled 'exact weighing, measuring and computing'.[38] This in turn imposed exactitude on the whole mining operation. Now, laboratories, assayers and chemists were commonplace, and increased specialisation of function had taken over from the mine manager, miner, metallurgist and businessman all rolled into one. And the availability of a *measure* of efficiency contrasted with the old days when the mine manager knew how much gold he had saved but not how much he had lost.

Australian goldfields saw similar changes. In evidence given to the Victorian royal commission on goldmining during 1889–91, numerous witnesses deplored the ignorance of gold losses, while Thomas Davey deplored the lack of laboratories.[39] Within a decade, assay offices and laboratories were no longer rare, and mining companies commonly appointed their own cyanide managers or metallurgists. In 1900 J. Malcolm MacLaren, B.Sc., commented that at Charters Towers the absence of assayers before 1895 concealed the high losses of gold in milling. Now those without their own laboratory facilities could take advantage of the services of numerous consultant assayers and metallurgists. By then it was clear that earlier cyanide treatments by 'labourers and miners, who were absolutely devoid of all chemical knowledge and worked by empirical rules', had left enough gold behind in the sands to justify cyanide treatment sometimes 'three times over'.[40]

That there were now people there to right this wrong was the result of cyanide's second route to the goldfields. Pushing the new technology into the structures of knowledge were those scattered scientists who from the early 1890s sensed its promise and sent future students onto the mining fields equipped with expertise in its ways. Professor Alfred Mica Smith at Ballarat had judged from his own small-scale experiments that cyanide extraction should form part of the metallurgy and assaying curricula of 1892.[41] By 1895 students could use a percolation plant and by 1898 they could work Deeble's agitation method. Diploma students spent at least three months as full-time assistants in the metallurgical works, and their duties included various cyaniding operations. Later, a specialised course on cyanide extraction was one of five programs devoted to particular industries. By 1902, twenty-seven people had received certificates of competency in assaying and metallurgy since the introduction of cyanide technology into the course, nineteen finishing in 1898 or later.[42] Of these, eleven received the full associate diploma.

At the South Australian School of Mines, chemistry examiners Professor Edward Rennie and A. J. Higgin expected student knowledge of the

Figure 12.2   Professor Alfred Mica Smith (*Australian Mining Standard*, 24 August 1899)

cyanide process in 1894,[43] and probably in 1893. It is likely that Rennie's advanced chemistry classes at the university included cyanide chemistry from the same time. At the University of Melbourne, where Professor Orme Masson lectured in metallurgy as well as chemistry, the cyanide process entered the curriculum in 1893.[44]

At Sydney University's new School of Mines, a test of cyanide competence was behind the 1893 exam question asking how to treat a quartz containing 2 per cent iron pyrites and 12 dwt of gold, but yielding only 4 dwt by amalgamation. In 1894 a detailed question asked about cyaniding losses and the technical advances which could minimise them.[45] In the same year, the school's Professor Liversidge was advocating the use of the process on the basis of his own laboratory experiments.[46] The following year, one of his students was investigating the factors 'inimicable to cyanidation' and its economic use.[47]

Students at Sydney Technical College were held back by the bias of William Dixon, whose syllabus did not refer to cyanide until 1897.[48] But by the time the Charters Towers School of Mines finally opened in 1902, the inclusion of the cyanide process in any mining course was obligatory.[49] Likewise, when the Victorian Government Metallurgist was asked to design a laboratory for research on the use of local mineral products, the inclusion of cyanide vats and filter press was standard procedure.[50] By 1903, James Park's 1897 pioneering textbook on cyanide had reached its third New Zealand edition. A year later came an English edition, a copy of which was obtained by Trail's Cyanide Works, Charter Towers.[51] By this time it was possible to choose from a range of cyanide texts.

As the cyanide process was establishing intellectual roots, students were adding another process to the armoury of skills and technical options they would take to the goldfields, and Australian mining was developing technological capability. James Thomas Dixon took his Bachelor of Engineering in Mining and Metallurgy from Sydney University School of Mines straight to an assaying position at Ravenswood, then in 1896 became chemist and cyanide manager to the Day Dawn company in Charters Towers.[52] Henry Leggo took his Ballarat competency certificate to Kalgoorlie's Kalgurli Gold Mines in 1901. Frederick Brinsden received his certificate in 1902 and was soon experimenting with western sulphotellurides.[53]

These people carried the influence of cyanide into wider realms. For perhaps its greatest impact lay with its demonstration effect, as a highly significant part of the process through which the application of scientific methods showed that mining was not so much about the rich strike but about the regular and efficient processing of the much more abundant low-grade ores. And nowhere was the benefit of cyanide and expertise more visible than on the famous and productive goldfields of Western Australia at the turn of the century.

Western Australia: triumph of the experts

Shareholders and directors of Kalgoorlie companies had developed high expectations by the time cyanide trickled through their more refractory ores. If its extraction performance was disappointing they would put it in the hands of those who could make it do better. Foreign-owned companies were ready to despatch experts to the colony if necessary, but chemists, metallurgists and engineers flocked to Western Australia, bringing 'the trained experience of the world' to bear on local problems.[54] The *Australian Mining Standard* observed that here Australian mining men of the old school, 'relying on their rule of thumb experience', were being left far behind their scientifically-equipped competitors, who were 'chosen to fill every important position in the great mining centres'.[55]

Julian Boyd Aarons was typical of those who came. Having studied mining and metallurgy at the Royal School of Mines in London from 1895 to 1899, he spent two years at Kalgoorlie gaining the experience which would secure him the management of a goldmining operation in Rhodesia. From there he returned to Western Australia to become manager of the White Feather Main Reefs, Boulder Deep Levels, the Vivien Gold Mining Co., then Chaffers Gold Mining Co. at Boulder.[56]

The Australian contingent among the experts came from the local elite: at least 30 per cent of the 1901 membership of the Australasian Institute of Mining Engineers had by 1903 brought their dedication to scientific mining to the west.[57] More than 20 per cent of the mining company representatives making up the Chamber of Mines at Kalgoorlie were members of the institute or had some kind of scientific training at an Australian school of mines, or both.[58] Some were significant figures in the Kalgoorlie mining community. John Waters Sutherland held an 1889 certificate of competency in assaying and metallurgy from the Ballarat School of Mines when, as metallurgist for Lake View Consols, he modified the filter press for Kalgoorlie cyanide practice. Within a short time he was manager of the Golden Horseshoe Estates and vice-president of the Chamber of Mines. At the Golden Horseshoe he applied his expertise to achieve 90 per cent recoveries, 40 per cent of which came from cyanide. In 1904 he was awarded the Associate Diploma in Metallurgy from his old Ballarat school.[59]

Sutherland was in good company.[60] One of several Ballarat contemporaries, Francis Ambrose Moss set up his own cyanide custom works, developed a unique and efficient recovery plant at Kalgoorlie, and whittled roasting costs to 4s per ton. He also became vice-president of the Chamber of Mines and president of the Institute of Mining Engineers. George Roberts was by 1899 assistant manager and metallurgist at the Great Boulder where he experimented with roasting before cyaniding and reduced working costs by 7s per ton. He suggested and supervised introduction of the Dhiel process at Lake View Consols and in 1904, as manager of Associated Northern, co-designed its plant. Another Ballarat man, John Dunstan, transformed the plant of Associated Gold Mining as metallurgist for this major producer.

Figure 12.3   The assaying room, Stawell School of Mines (*Australian Mining Standard*, 3 March 1898)

The flow from the Stawell School of Mines included A. E. Bennett, assayer at the Coolgardie Cyanide Co., and A. G. Brown, assayer at Bayley's Reward.[61] Among the university graduates, Sydney's E. W. Nardin took control of Hannan's Star bromocyanide plant in 1904 and introduced several procedural modifications. Hugh Burton Corbin went from the University of Adelaide to become an assayer at Kalgoorlie from 1894 to 1898 before becoming manager of the Arltunga government cyanide works back in his native colony.[62]

Others learned from experience that this high-profile western market attracted and rewarded those who had put themselves to the trouble and expense of acquiring some scientific education. In 1897 H. Swan Edgar withdrew from practical mining in Western Australia to gain professional training at Ballarat. He returned to the west as a cyanide manager from 1902 to 1905, and then as a goldfields consultant.[63]

In Western Australia the energetic and innovative adaptation of extraction processes was crucial to the successful exploitation of deeper layers of wealth, a lesson not lost on members and observers of the Australian mining community at large. The *Australian Mining Standard* articulated the message in 1901:

success no longer depends on rich returns, but upon the steady average from the low-grade ore. The metallurgist, the geologist, the chemist, and the engineer are the men to whom the great expansion of this industrial field is due.[64]

Scientists and educational models

Some scientists had been pressing this kind of point at least a decade before. Robert Jack's message in Queensland resonated with the educational model espoused through the forum of the Victorian royal commission on goldmining. When Alfred Mica Smith described a structure where three- or four-year courses in mining engineering, metallurgy or geology would lead to Associateship of the Ballarat School, he sought to shift mining education out of the random informality of the cultural infrastructure and into its own stratum of systematic and specialised professional training.[65] With the mining commission's endorsement, the Ballarat School of Mines moved in 1891 to begin adding a second tier to mining education.[66]

In South Australia the scientists on the Technical Education Board had culled ideas from Victoria and the world,[67] and were building a similar model into the South Australian School of Mines and Industries, which opened in 1889. Three-year diploma programs in six subjects devoted at least one year to one specialisation.[68] Under a special arrangement, students attended lectures at the University of Adelaide in advanced chemistry, physics, geology and electrical engineering.[69] Lesser aspirations were catered for with certificates to successful candidates at annual examinations. But no candidate was allowed to sit examinations without a satisfactory report of conduct and 80 per cent attendance at lectures and practical work.

In New South Wales, the model of mining education was extended further when the 1892 deliberations of the Government Geologist and Professors Archibald Liversidge and Edgeworth David added a third tier to institutional and qualification structure. An integrated system now linked the technical college branches in mining centres, the Sydney Technical College, the university's engineering faculty and a discrete and specialised School of Mines, located within the university but affiliated with the Department of Mines. Specialist mining degrees were now an option where before the ultimate had been a diploma. The university provided the Bachelor of Engineering in Mining and Metallurgy, while the Technical College provided the structure for a Mining Managers' Certificate. By 1900 there were eighty students enrolled at the School of Mines, and many graduates were in receipt of salaries over £300 per year.[70]

But still there was discrepancy between scientists' models and the 1890s reality of perceptions and recruitment on the fields. In Victoria, royal commission witnesses had overwhelmingly favoured undemanding supplementation of practical mining knowledge on locale.[71] In South Australia, climbing student numbers paralleled a tendency for students 'to take up subjects as occasional courses'.[72] In Queensland an 1895 attempt to form a school of mines at Charters Towers failed for the same reason as in 1888, and the limited range of subjects available at the Workers' College was no substitute. Even support from the local Chamber of Commerce and Mines in 1897 was tinged with uncertainty as to whether a school of mines would attract many students, a doubt validated by the opinions of witnesses to the 1897 royal commission on goldmining.[73] The parallel Western Australian royal commission found a balance of opinion in favour of schools of mines, but split between those who espoused the *ad hoc* model of the cultural infrastructure, and a significant number who were quite definite that something better was necessary.[74]

Neither commission led to strong recommendations or government action. Yet a couple of years would make a difference. As Queensland's cyanide gold production peaked, goldfields representatives were eager to redress the observed mismatch between the demands of refractory and low-grade ores and the available expertise. South African experience with the cyanide process was cited as evidence of what was possible.[75] When Robert Philp agreed as Minister for Mines to discard the previous policy of leaving the initiative to local communities,[76] the way was cleared for the 1902 opening of the state-funded and -administered Charters Towers School of Mines, 'untrammelled . . . with technical and trade classes'.[77] Western Australia was moving in a similar direction. Under the pressure of intense and unified lobbying from the Chamber of Mines and the mining union, a school of mines was founded under the auspices of the Department of Mines in late 1902.[78] In 1904 it moved from Coolgardie to a new building at Kalgoorlie.

The Charters Towers and Kalgoorlie schools both represented the new ideals of professional scientific mining education. The primary objects of 'thorough training' for students and direct benefit to the mining industry were supported by appointment of specialist, credentialled staff. Courses were well organised, systematic and regularly reviewed; laboratories were well equipped and high standards were guaranteed by respected outside examiners.[79] The Kalgoorlie mining community expressed its support through providing scholarships, work experience and employment. This was reciprocated with future research strongly linked to practices on the field.[80]

By early in the twentieth century, as Adelaide and Melbourne universities followed Sydney's lead with specific provision for mining students,[81] a new model of mining education was identifiable in *practice*. It consisted of three different institutional and qualification levels, corresponding with different degrees of scientific knowledge and skill required within the industry.

- At the lowest level the cultural infrastructure still provided some 'useful' knowledge. Through miners' institutes, schools of arts, mechanics' institutes, and small schools of mines, local on-the-spot access to knowledge provided in an unsystematic and undisciplined way was still available to mining communities.
- At the next level schools of mines and industries, and some technical colleges, provided a more systematic program of training for a particular qualification. These ranged from stage certificates to certificates of competency to three- or four-year associateship diplomas in the most professionally oriented institutions. At the upper end of these, the Charters Towers and Kalgoorlie Schools had set a new style by catering for mining students only.
- At the highest level were the centralised schools of mines for top-level scientific study, attached to universities and staffed by university scientists. Entrance requirements and disciplines of attendance and progression were prerequisites to the achievement of a degree and full professional status. Here research had an acknowledged place.

A model which had grown out of the first level featured an integration which allowed upward progression through the levels of education and thus through a mining career. This linkage was maintained as the developments of the 1890s defined a specialised mining niche and progressively upgraded its quality.

Though implementation was imperfect, the bulk of mining education had been shifted from the first level, where it lay at the end of the 1880s, to the second level, where systematic courses of instruction were the order of the day. As the Victorian Government Metallurgist noted in 1905, the day had passed 'when the teaching of metallurgy was considered finished after some quasi-chemical lectures and a course of assaying had been added to a more or less lengthy acquaintance with the strictly empirical practice followed in some works'.[82] In addition, the elite learning provided within the universities had been made relevant to the needs of mining and gained credibility on the goldfields. The work of Jarman and Brereton at the University of Sydney struck a previously uncommon note in mining science. Their research programs aimed at one of the remaining critical problems for the cyanide process, investigating the use of ammonia and its compounds in cyaniding copper-containing ores and tailings.[83]

The link between mining performance and the application of scientific expertise had for some time been clear to those with technical training, but it took tangible evidence to push the industry to demand facilities for the scientific training of Australian mining men and to push governments toward their 'manifest duty' to provide them.[84] The cyanide process had been a significant part of the end-of-century mining transformation which shifted the climate of opinion. That shift in attitudes and capability is evident in the membership of the Australasian Institute of Mining Engineers. From small beginnings in 1893, those who were explicitly dedicated to the application of science to mining and met the required standards of qualification and experience, had expanded by 1912 to a total of 438 ordinary members and 233 associate or student members. Of the total number of 671, only 61 had been elected before 1900.[85]

Cyanide gold extraction: reflections

The cyanide process was steered from its centre in Glasgow, a British technology made for transfer to the major gold-producing countries of the world. In Australia, geographic conditions effectively determined that this be done by sale of licences. This meant its successful exploitation depended on widespread assimilation into Australian goldmining. Though several components of a foreign technological system were transferred, such as capital, experts, machinery, cyanide and organisational structure, there could be no isolated foreign enclave.

Despite the advantage of many colonial connections, neither the transfer itself nor the subsequent process of diffusion was simple or inevitable. Australian gold production occurred through a collection of colonial sub-systems, each to be dealt with separately according to its own features of economy, legislature, ore bodies, people and environmental conditions. While they had much in common, they had significant differences.

Diffusion initiated from three different colonial centres faced divergent treatment problems, in terms of the ores themselves and of the standard practice which would have to be broken down. The years up to 1897 brought little encouragement as obstacles ranging from geography, geology and ignorance to depression, legislation and alliances of vested interests combined to keep the process at bay.

But take hold it eventually did. For the Australian gold production system also had some favourable features. Through various mechanisms, governments aided the articulation and communication of problems as well as a flow of expertise and institutional change. A national mining journal spread mining information across geographical and political boundaries, and consistently promoted the use of scientific methods in mining. A core of professional mining men who had banded together in 1893 as the Australasian Institute of Mining Engineers were dedicated to breaking down the divisions of distance and conservatism and promoting the improvement of mining methods, including the cyanide process. All these factors aided adaptation to a technology which was by no means amenable to transfer by 'blueprint'.

The immediate effect was to increase the gold yield by significant amounts. Even when the process was applied only to the residue of tailings after amalgamation and/or roasting, the extra gold recovered often made the difference between working an ore payably or abandoning it. In addition, the possibility for coarser crushing increased milling throughput and reduced costs. How much of the extra wealth stayed in Australia is questionable. Since the majority of Western Australian gold companies were foreign, much of the extra dividends would have been repatriated back 'home'.

The Australian gold production system entered the 1890s highly dependent on Britain for most of its methods and associated expertise. And it was becoming more, rather than less, dependent on Britain for the increasingly large capital funds required for large-scale and more complex mining and processing. On the other hand, diffusion by licence held some potential for real transfer of knowledge and skill, both through educational channels and through learning by doing on the fields.

In the 1880s, certificates of competence represented the highest level of specialised mining qualification, an expertise held by a very limited number of people. By the time the cyanide process had become an established part of Australian mining practice, the industry's technological capability had risen. Imported experts might come and go but now there were specialist mining institutions and qualified people, and they were much more likely to find employment and credibility.

The companies which had brought the process to Australia had also undergone some change. The AGRC reconciled itself to the lost potential for winning fortunes from licensing the use of cyanide. Early in the new century the company was liquidated and reconstructed as the Australian Mining and Gold Recovery Co., in recognition of the fact that a more certain future lay in running its own mining and extracting operations. The Cassel Gold Extracting Co. had also recognised that its future lay with production of the cyanide that was now being used on goldfields around the world. In 1906 its name was changed to the Cassel Cyanide Co. In 1927 it was absorbed into the merger of major chemical companies which formed the giant of the future, Imperial Chemical Industries (ICI).

PART IV
Linkages, learning and sovereignty

CHAPTER THIRTEEN

Transfer, diffusion and learning

Understanding of technical change in its various forms moves on. Technology is more than machines and techniques, the technology gap is more than the mere lack of them, and the development gap between countries is more than a technology gap. Theorists tell us that economic development is a process characterised by the interplay of two conflicting forces: innovation tends to increase economic and technological differences between countries, while imitation and transfer tends to reduce them.[1] But what is the relationship between these two forces; and can technology transfer generate a capacity for innovation?

The industrialisation of Western Europe showed how the acquisition of Britain's technology triggered a process in Continental neighbours which brought them ever closer to the industrial leader. Yet, as the analytical focus moved on to a new century and other locations, the evidence seemed to reveal an opposite trend, where those who possessed the new wealth-generating technologies could parcel them out in pieces so as to retain control and most of the gains. Now neither process seems so certain.[2]

In fact, the specific factors which bear on the outcome of technology transfer are numerous, as borne out by the cases of anthrax vaccination and cyanide extraction. In the broadest terms, we can look to the disparity between the imported technology and the local system with which it has to find some kind of technical and cultural fit. Secondly, the characteristics of the receiver are vital, in particular, its capacity to diffuse the imported technology, its capacity to capture and distribute the potential benefits, and its capacity to respond and change. These may be affected by the relationship with the sender.

The technology gap

The technology gap has several dimensions. The technological systems approach which leads us to conceptualise this void in terms of the discrepancies between transferring and receiving systems takes in cultural, institutional and resource differences. At a more purely technical level, however, it must also include what has been designated the scientific and technical knowledge gap, and, on another level, an experience and skills gap.[3]

Costs are always involved in adopting new technology. These include the obvious, such as equipment investment, licence costs and the like. Less obvious are the information and screening costs involved in seeking and assessing the alternatives, the cost of time spent by instructors, supervisors and workers in learning how to use the new technology efficiently, and the costs of mobilising skills in management, marketing and distribution. Obstacles to movement toward the technological frontier may be overcome, but only at a price which reflects its distance.

For changes in technology can be dramatic. While many new technologies evolve out of improvements to those already in use, others involve a radical shift to new concepts, new sets of knowledge, new combinations of techniques, skills and organisational and institutional arrangements. There is, in the latter case, a distinct *discontinuity* between new and old, and the cost of acquiring new facilities, knowledge and experience may be great.[4] Firms, industries and countries can become 'locked-in' to inferior technologies through commitment of resources along a specific technological trajectory.[5]

In 1880s Australia, the system for controlling stock disease recognised the phenomenon of contagion, but had not yet caught up with the metropolitan verification of living organisms as the agent of disease. The diagnosis of Cumberland disease, and the standard practice for its management, shared little in the way of common knowledge, skill and experience with the theory and practice of attenuation and vaccination. Added to this were those 'environmental' factors arising from the 'location' gap.[6] In particular, the vaccine was designed for small French farmyards not Australian broadacres, and it was accompanied by a cultural rift, part of which was manifested in the difference in language, and part in national stereotyping. Both were compounded by the difficulties of communicating from the other side of the world.

The transfer of cyanide gold extraction benefited from the sharing of common institutions, language and culture and from the ready flow of people. A multitude of colonial connections were able to smooth away early obstacles in the patent system. More difficult was the distance between the practices of Australian goldmining and the leading edge of the technology frontier which this chemical process represented: despite some pockets of experiment with chlorination, the vast bulk of the goldmining industry was ignorant of chemical extractive processes and the scientific principles which governed them. These were considered the province of the much maligned 'theorists'.

Yet colonial Australia showed an ability to change course, to cross the divide between old and new trajectories. Discrepancies between systems proved surmountable, the technologies were transferred and adopted widely in the relevant sectors of industry. In neither case was adoption instantaneous: analysis of the diffusion process showed it took some years to reach its peak. In seeking the conditions that laid the basis for acceptance, we confront another factor bearing on the outcome of technology transfer—the features of the receiving technological system.

The capacity to diffuse

Rosenberg has argued that 'from the point of view of world history the critical social process requiring examination is that of diffusion':[7] the receptivity to new technologies and the capacity to assimilate them whatever their origin are 'as important as inventiveness itself'.[8] Many different kinds of factors have been identified as influences on the rate and extent of adoption of new technology—economic, technological, institutional, socio-cultural, geographic. The 'epidemic' models which underlie most diffusion studies, however, rely on an analogy between the spread of innovation and the spread of disease, and emphasise individual propensities for 'infection'. The explanation of how warmly a new technology is received within a given population then lies with demand factors, especially the relative distribution of risk-takers and risk-avoiders.[9]

Such shortcomings have recently attracted many criticisms and new attention to the supply side of diffusion.[10] It may, for instance, be quite rational to await the technical and other refinements which often improve the price and quality of new technology. However, as Stoneman has argued,[11] if it is considered that what is being spread in epidemic fashion is information, not only about the existence of a new technology but also about its performance in practice, then the result is a kind of learning, which thereby reduces uncertainty and risk, factors which have been found to be substantial barriers to adoption of new technology.[12] The increasing recognition of the significance of formal and informal networks in the spread of relevant technical information[13] confirms the importance of those factors which affect information flows.

For the diffusion of new technology in a vast, unpeopled and undeveloped country such as nineteenth-century Australia, information flow was obviously critical. Here two features of these Australian systems stood out. One was the significance of the administrative networks erected by governments; the other was the appearance of national industry journals and associations.

The network of sheep districts and stock boards carried the news of the Pasteur vaccine and then allowed Gunn to extend his market outward from the Yalgogrin property. As an insider with substantial credibility, Gunn exploited his minimal *social* distance from potential users, and thereby minimised the *geographical* distance, helped by good press in a pastoral journal which monitored publications and events in all the colonies. The 'neighbourhood effect' which was apparent in the early use of the Australian vaccine was therefore gradually overtaken by other ways of carrying information.

The administration of mining districts was confined to smaller, more isolated areas, but channelled information to and from a central bureaucracy through mining wardens and publications. Though most miners might never see a departmental report, the national mining journal provided regular summaries. This supplemented its commitment to report the latest in mining science and technology. At a more elevated level, a national institute of mining engineers aimed to unify and inform the industry's professionals.

Diffusion and adaptation

But knowledge of the existence of technology was not enough. Various kinds of adjustments were needed to reduce the dissonance between new and old, foreign and local. The first requirement was for some means of translating between the two; here key individuals in the infrastructure which serviced each industry encouraged common definitions of the target problem, leading to common perceptions of possible solutions. Transfer agents and potential users could begin to talk the same language, inhabit the same conceptual world.

Then the improvers went to work, adapting foreign technologies to a new environment. It was as a self-taught user of anthrax vaccine that Gunn combined with a local scientist to mould the product to Australia's climate and geography, and to its labour and industry structure. The result was four sequential innovations vital to widespread diffusion of the product. With cyanide, Cassel experts carried out some modifications, but others were achieved by local users, both big and small, and by scientists, both government and private. Limitations were evident in the small number of Cassel experts, and in the local supremacy of the 'rule of thumb' men and the weight of untrained prejudice and habit.

Institutional adjustments also smoothed the assimilation of imported technologies into the Australian landscape. A new law provided government control of the introduction of disease micro-organisms; a government policy which would not countenance lump-sum purchase of a 'blueprint' would encourage establishment of a vaccination sub-system in Australia with indirect forms of assistance. The government also resisted repeated lobbying for compulsory vaccination and for state takeover of the McGarvie Smith and Gunn manufacture. But eventually, their methods were incorporated into separate institutions, one entirely under state control and one under a trust guarded by government and pastoral representatives.

In the case of cyanide, observable jumps in Victorian usage coincided with specific government moves to stimulate the mining industry. Other institutional responses included government metallurgical works and consultancy services, and the passage of legislation to prevent renewal of the cyanide patent. There were also concerted campaigns to cancel institutional support for the royalty: some were overtly economic, others co-opted scientists into a dialogue with the legal system. The overall effect was to reduce the cost of crossing the technology gap and taking up a technology on the frontier.

The benefits: the link from foreign technology to domestic economy

Both these cases of technology transfer could be said to have been 'successful'. Both were assimilated into local production, bringing immediate benefits as they did so. According to J. W. McCarty, one of the benefits of cyanide was a revival of goldmining from the late 1880s due to the hitherto unprofitable deposits brought into production.[14] This technology, he said, helped lay the foundations for economic revival in the 1890s, a view which revised the traditional significance given to the 'extensive' bonuses from new-found gold deposits. Blainey's subsequent warning[15] against attributing too much to a process which was little used before 1894 and from then only added small additional percentages of gold from discarded tailings may, however, have underestimated the overall effect.

We can quibble about specific direct effects: Blainey suggested that even in the peak year of 1903 cyanide probably added only 10 per cent to Australia's gold production; it has been suggested here that it may have been at least 20 per cent. More important is the fact that the cyanide process made economic contributions in a number of ways, both direct and indirect, both extensive and intensive. Cyanide did not simply 'add on' extra gold production through its application to tailings which would otherwise have been discarded. This was an important dimension of its use, but not the only one.

It has been suggested that incremental innovation may be more economically significant than radical technological change.[16] Certainly incremental innovation is the mechanism which ties transferred technology more firmly into the local economy. Once we see that the diffusion of cyanide extraction was not simply the increasing adoption of a single innovation, but a *process* of cumulative adaptations, it becomes clear that there are several sides to its economic impact. Gold output was augmented directly through the treatment of both old and new tailings which would otherwise have been left untreated. Innovations designed to cope with the very finest portion of the tailings extended the proportion to which it could be applied. In the 'all-sliming' methods of Western Australia, cyanide was often effectively applied to the whole of the sulphide ore through the whole process, even though amalgamation might still be used to take out any coarser gold in a separate step. Thus adaptations increased the extent of the ore on which the process could be used.

Small-scale innovation was also lifting the efficiency of *using* the process. It not only helped add on extra ounces of gold, but did so at steadily reducing *cost*. Productivity gains derived from extra ounces of gold per ton of treated ore, lower costs per ounce of recovered gold, and extra ounces of gold per labour input for a given ore. Cyanide made a contribution to all of these, both directly and through cheapening other parts of the extractive process: with less reliance on physical separation, coarser milling could boost battery throughput, while the requirement for accuracy and measurement implied monitoring of all operations. All these productivity improvements made it possible to work deposits not previously payable, and indeed changed the approach to the mining of low-grade ores. There was therefore an interaction between the extensive and intensive aspects of growth. The latter came from the steady accrual of productivity benefits through small changes, technical and institutional, artefactual and organisational, made along the way.

These points emerge obviously for cyanide; they can also be made, perhaps less obviously, for anthrax vaccine. As a preventive technology, its economic effects are less quantifiable, but wool production was boosted by eliminating a cause of stock losses, and having more sheep in active production. Discarding the practice of taking suspect fields out of use had obvious productivity implications, particularly at a time when drought was decimating both flocks and pastures. These benefits were intensified, and extended to a greater proportion of threatened pastor-alists, as the price was reduced by a series of innovations.

Economic benefits and technological capability

Clearly, more than static benefits were delivered by these overseas tech-nologies. The continuing process of adaptation made them viable for a widening circle of users, and improved the gains to individual users over time. It is for such reasons that some writers on diffusion have made the distinction between adoption and 'absorption', the process which brings the full potential of new technology to individual firms and the industry.[17]

Contrary to earlier assumptions, different levels of risk aversion are not all that affect technological decision-making. Since the work of David,[18] it has been recognised that specific features, such as firm size, affect the cost of adoption and hence the threshold of acceptability, a factor evident in the pause before anthrax vaccine fell to a price level accept-able to small producers. Advocates of recent evolutionary models of tech-nological and economic change have extended this line of thought to argue that decision-making occurs within the context of a bounded rationality which varies for *all* adopters. Firms need to learn the eco-nomic properties of a new technology, not just in some general sense, but in terms of the precise relevance to their own particular circum-stances. There is a distribution of costs and benefits of adoption; infor-mation requirements and assessments vary, as do the criteria of assessment and the economic and cognitive capacity to perform it.[19]

These factors suggest that the tendency to adopt new technology will depend on the ability to scan, assess and select technology that is rele-vant to user needs. Indeed, without some such complementary research capacity, technology transfer is unlikely to take place. Knowledge and technology are not costlessly available to all-comers: 'it takes a research capability to understand, interpret, and appraise knowledge that is placed on the shelf—whether basic or applied'.[20] This must be especially so for leaps across technology gaps toward the technology frontier. It is for this reason that Fagerberg has argued that economic growth depends on three factors, not two; to technology transfer and domestic innovation has been added the development of internal capacity for exploiting the benefits of available technology, regardless of source.[21]

There are several kinds of relevant domestic technological activity and associated capability. Those expressed by a range of authors[22] distil into five types: searching and screening; implementing particular technolo-gies; adaptation at various levels; undertaking comprehensive research; and dissemination. The lesson has been drawn that to catch up with the

developed countries, the less developed cannot rely only on a combination of technology import and investments, but have to increase their
national technological activities as well.[23] The effectiveness of the link
from foreign import to domestic integration depends on domestic capability. This begs the question of how this attribute is acquired in a
country forced to import capability in the form of foreign technology.

In Chapter 1 reference was made to the suggestion that a progressive
process of technology transfer can effect a transition from dependence
on borrowed technology to the capacity to invest in indigenous technology. From a need to import a finished technological product, it is
possible to move to importing 'blueprint' designs and then only the
general knowledge base. This implies that technology transfer brings
some process of learning which elevates domestic abilities. How is this
learning likely to come about?

Learning benefits and technological capability

Much has been made of Arrow's original observation of the significance
of 'learning by doing', the development of on-the-job skills by operators
in manufacturing, and the associated productivity gains.[24] Rosenberg has
also drawn attention to the scope for 'learning by using', where users
modify technology to their specific needs.[25] This adds knowledge to the
development process, brings the user to a higher level of skill, and may
feed back into product design, giving rise to improvements for all users.

Bell has warned, however, against assuming that performance
improvements are *caused by* some accumulation of knowledge and
skills: to do so would imply that doing-based learning arises passively,
virtually automatically, as a costless by-product from carrying on production. Instead, Bell distinguishes between two different kinds of learning by doing—those that come 'free' and those which depend on the
allocation of effort and resources to capture the learning potential. The
limits of the former tend to be reached very quickly. It is the latter
variety which boost technological capacity in meaningful ways.[26]

There is plenty of evidence to show why this is so. Built-in feedback
systems can provide an understanding of how and why performance
varies, but require the instigation of monitoring programs and sufficient
skill to interpret the information obtained. Formal training, perhaps as
part of a technology transfer package, has also been found to be effective,
while conversely, the lack of such programs has been found detrimental
to efficient use of transferred technology. Firms can, of course, hire skills
ready-made, but this depends on their availability and the allocation of
funds to acquire them. Even 'disembodied' knowledge will come through

a succession of costly technology contracts with foreign consultants, producers and suppliers.

If Bell's interpretation of the evidence is correct, doing-based learning may provide productivity benefits and technological experience within a given production process, but does not, of itself, lead on to acquisition of other types of technological capacity. In particular, it will not carry a firm into the realm of a significantly different technology. This requires the deliberate instigation of other mechanisms for learning. In this regard, *institutional* investment, development and innovation become critical, determining where and how the learning takes place and whether the learning benefits filter through to people in the receiving country. In the words of Perez,

> Previous capital is needed to produce new capital, previous knowledge is needed to absorb new knowledge, skills must be available to acquire new skills and a certain level of development is required to create the externalities that make development possible.[27]

In the case of anthrax vaccine and cyanide, users faced a discontinuity, a shift into a more sophisticated technology based on a body of scientific knowledge and techniques. This in itself seemed to offer the potential for learning. Yet it meant that the opposing tendency—to become 'locked in' to an earlier technological trajectory—would have to be countered, unaided by tacit knowledge absorbed from former processes. The result revealed that certain kinds of learning were taking place, but none flowed automatically or without effort or cost. They depended on the existence of some kind of relevant local capability and on the actual nature of the transfer mechanism.

It was a reservoir of bacteriological knowledge in the persons of Willows and Stanley which made possible the identification of the anthrax problem and pointed to a new solution. Though Stanley and Hamlet failed to implement anthrax vaccination from the 'blueprint', their own familiarity with vaccine technology, enhanced in the process of acquiring information from overseas, was passed on through what was effectively a process of informal training as they introduced the concepts, terminology and prevailing theories to Bruce and Devlin, who, in turn helped spread them to stockowners through the pre-existing networks of the government-imposed stock boards.

This French system of vaccination did, however, seem biased against further learning, the usual practice being preparation of the vaccine at the Pasteur Institute in Paris and despatch of the finished product to the desired location for injection. Australia was fortunate that distance offered opportunity when the risk of deterioration during the long voyage made necessary the establishment of a local 'substation' to carry out the final stage of manufacture. French representatives still prepared and dispensed the final product: users merely paid the cash and looked for the benefits, ignorant of the inherent technology. But in practice, the location of the French in Australia made possible the 'leakage' of information through personal relationships.

The use of an Australian vaccinator arose from the previous close association between Devlin and the Pasteur representatives, and further strengthened that bond, in the course of which Devlin and his sons learned a great deal about bacteriology, anthrax and vaccination. A later expert concluded from Arthur Devlin's 1888 paper on anthrax that he had 'mastered his subject theoretically at least as regards bacteriological techniques and practically as regards the various features of the disease as it affects animals'.[28] Though the Devlins were not transformed into professional bacteriologists, their later careers as stock inspectors injected into local production a level of familiarity with theory and practice that went beyond the experience accrued in injecting the vaccine into animals. It reflected instead the informal 'training' they were given by the French.

Users underwent one level of learning as, through experience in balancing risk with benefit, they were able to optimise their use of vaccination technology. But with little need for them to master knowledge or techniques, the impetus for further learning came from adaptation of the technology to its new environment. Here the departmental collection of performance statistics provided the feedback which signalled that changes needed to be made. The Frenchmen made a concession to climate by changing vaccinating times to cooler months, but it was the presence of an intelligent, energetic user, himself monitoring performance of the technology in his local area, which led to adaptation of the product itself. This supports Lundvall's contention that interaction between users and producers is another element of technological learning and that the scope for such is a vital ingredient in innovation.[29]

Having detected a problem, John Gunn and John McGarvie Smith passed through a considerable period of learning, developing their bacteriological expertise as they looked for an answer to the unpredictability of Pasteur's vaccine. It was a process which relied on their commitment to use any relevant resources and networks, and to endure the rigours of a lengthy learning curve before they were in a position to begin modifying the product. Formal training was not yet available, but both used informal channels to acquire general bacteriological knowledge and specific aspects of anthrax vaccine technology. Such sources were present in a strengthening local scientific community with ready access to the scientific culture of the metropolis. Personal contact with the French added the opportunity to observe the 'uncodified' aspects of the technology. As knowledge and technique were mastered, the Australians progressed up the scale of technological activities and capabilities. Users became imitators, then adaptors. For a time they engaged in research aimed at deeper understanding.

In the winning of gold from Australian ores, an emerging awareness of deficient capability ran in parallel with a trend to expend time, effort and funds in extensive searching of the international technological shelf for an answer to the goldmining 'problem'. Though preceded by many mining agents, George Fowler and William Burrall were the first Australians to commit themselves to the cyanide process as their chosen technology. Burrall was helped by his background as a mining engineer.

Again, adoption of this new technology generally meant crossing a technological discontinuity, from physical separation to chemical extraction. This made past experience somewhat irrelevant. Yet the nature of the technology, and the way it was transferred, called for a certain competence on the part of the user, a certain mastery of the techniques and knowledge which were required to make the process work on a given ore, in a given environment.

Again, distance from the technology source played a part. Transport costs ruled out the shipment of the ores for treatment in Britain. Even within Australia, transport costs within such a dispersed industry ruled out a central processing works for treatment by the British organisation. Therefore technology transfer came through formation of an Australian subsidiary structured to 'farm the royalties', rather than the ores. Though the company provided its own expertise to smooth the way of adoption, the ongoing business of bedding in and optimising the process depended on the availability of expertise at the mine. The primary role of the Cassel experts was therefore to 'train' cyanide managers and operators.

It was not an expertise which transferred automatically. George Fowler saw only too clearly the distinction between those who could learn to operate the process on a routine basis and those who could adapt the process to the ores. There was some capacity within local institutions to produce the latter, through schools of mines, and as the new technology was included in their curricula, students were trained in the theory and practice of the cyanide process. Mining companies could begin to hire the skills they required. The gap between need and availability was bridged by borrowing the expertise of government officers, and also by hiring from overseas. In many cases it was a matter of muddling through, but the fact that the tailings left *after* treatment could often undergo another two rounds of extraction when chemists were on the job showed what knowledge could do.

The British company was of course adding to that knowledge all the time, through its own research laboratory in Glasgow. But its attentions were directed world-wide. In Australia the range of conditions meant that much of the onus fell on the local system. Here there were both positive and negative responses, as the colonies exhibited different needs and capacities for adaptation and innovation. It was least evident in Queensland, where the process was characteristically one for 'the little man', and most evident in Western Australia, where it was absorbed by the highly capitalised foreign corporation.

In between, however, were several instances of adaptive activity which enhanced capability in the system. Goyder is perhaps the classic example: his screening of cyanide led to more efficient production monitoring and to adaptive research, which in turn led to investigations of the why and how. In the process he increased his own understanding and competence as well as that of the young graduate from the Adelaide School of Mines who was made manager of the Mount Torrens plant. Grayson took this experience when he moved onto greener pastures in Western Australia. Likewise, Mulholland's work had wider offshoots through his prominence in the Bathurst Science Society which he reinvigorated with his presidency and his papers on gold extraction.

Technology transfer and innovative capability

So what of the argument that technology transfer inhibits the development of local competence and innovation? It might perhaps be argued that, had the Pasteur vaccine not been imported, Stanley and Hamlet might have perfected their own technique of vaccine preparation. Stanley himself believed that the bacillus would have to be studied in Australia, under Australian conditions, but the secondment of Katz's expertise was pre-empted as the heralded arrival of Pasteur's representatives offered the easy way out. William Camac Wilkinson was also adamant that there were at least six men in the colony who could prepare a Pasteur-type vaccine, although there is no evidence that anyone was attempting to do so. The mysterious 'six' were probably medical doctors, like Wilkinson himself, with some knowledge of bacteriology but little interest in or time for pastoral problems. Much clearer is the fact that the need to adapt the Pasteur system to Australian conditions brought McGarvie Smith and Gunn into a field they might not otherwise have entered, and provided the motivation for them to develop their own competitive product.

Local attempts at chemical solutions to gold extraction problems included Dixon's early efforts, but essentially came down to modifications of chlorination by Newbery in Victoria, Hall and Richard at Mount Morgan, and Brown at Charters Towers. Otherwise, experimental activity focused on the screening of transferred techniques, most of which were directed toward the physical separation of the gold. As with anthrax vaccine, the publications of the scientific enterprise showed no evidence of any prior research programs directed toward the problem, apart from the work of Storer. Certainly no work was displaced. Instead, cyanide added to the options and had to compete on its merits. In so doing, it provided a centre around which a wider circle of experimental programs would cluster.

Rather than inhibiting local effort, technology transfer stimulated *complementary* innovation and learning, and led to higher levels of technical competence. The danger area was in Western Australia where an army of overseas experts flooded into foreign mining companies. Here was the potential for these transient experts to scoop up most of the learning potential, while their overseas companies scooped up most of the dividends. As the president of the Institution of Mining and Metallurgy in Britain pointed out to his colleagues, the mining industries of Britain had dwindled to such an extent that it was on colonial mining that British metallurgists depended to practise their profession.[30] The saving grace in the situation was the concentration of Australia's own mining vanguard: they competed successfully for prime positions, and participated in the innovation.

Again Bell's point is validated—only through the prior allocation of resources to learning are further substantial learning benefits likely to be captured. No doubt a greater investment in educational institutions would have yielded greater local benefits from Western Australia's boom. Yet the problem has to be seen within the context of a young immature nation that was allocating resources under the constraints of having to *grow* in broad dimensions as well as wealth.

Under this condition, it is possible to see a process through which the activities surrounding the entry and assimilation of overseas technology moved progressively into areas of greater complexity. Goyder progressed from screening to adaptation to comprehensive research, and Gunn progressed from using to monitoring to emulation, then to adaptation and beyond. And all the time, they were touching a widening circle of people who impinged on the conduct or transmission of their technological activity.

Perez has suggested that windows of technological opportunity may open for less developed countries at times of world transition to a new 'techno-economic paradigm'.[31] In these Australian cases, perhaps even

more long-term significance emerged from the very size of the technological gap that was encountered. In each case the qualitative difference
between old and new exposed the inadequacies in the local system, and
provided a demonstration effect of the benefits to be gained from the
adoption of 'scientific' methods, and from developing the expertise
which could wed science to the productive process. From such a climate
came an institutional response designed to elevate the technological
capability of each industry. It was not long after Loir established the
Pasteur Institute on Rodd Island that support emerged for a local institute
both for training local people in bacteriology and for conducting research
in animal disease. Similarly, growing demand for scientific mining education was informed by a more professional model than had characterised the demand of the 1880s. In the early years of the twentieth century,
the Australian states were analysing more generally their institutional
capacity to generate technical skills.[32]

Within new institutions, knowledge of new technologies was handed
on to a new generation of mining and veterinary students, bringing them
towards the frontier of established knowledge. As a receiver of imported
technology, Australia was better equipped; future technology transfers
would no doubt deliver potential benefits more effectively. But what of
the capacity to innovate?

Adaptive, incremental innovations established the imported technology but revealed nothing of the radical invention which can lead to the
birth of new technological systems. The problem here was the limits to
institutional change. While there were moves to upgrade technological
skills for the purpose of using modern technologies, there was no secure
and clearly allocated institutional niche for systematic and comprehensive research. The plans for a stock institute faded away: the eventual
outcome was a slot within the Microbiological Bureau, where limited
investigations were always subject to the constraints of a government
department; quite separately, a Chair of Veterinary Science catered for
the scientific training of veterinary students. Likewise, any research
carried out in conjunction with scientific mining education was wedged
between functions of higher priority. 'Breakthrough' research relied on
the extremely innovative individual, on the maverick in the system, or
outside the system, rather than on a systematic approach in defining,
understanding and solving the critical problems of production systems.

In this respect, a series of moves around 1916 were significant. The
formation of the Advisory Council of Science and Industry was an attempt
to create an institutional niche for research which was directed at the
critical problems of primary and secondary industry. The formation of
the Walter and Eliza Hall Foundation created a niche for immunology in
general. The formation of the Commonwealth Serum Laboratories

created a niche for the research and production of vaccine technologies. Though the consequences are another whole story, and the effectiveness of the results is still being debated, their creation marked a significant step forward in institutional response to the need for system change. In the meantime, colonial science had another role to play.

CHAPTER FOURTEEN

Colonial science: the intellectual bridge

The position of scientists at the periphery seems paradoxical. On the one hand they were remote and removed from the intellectual and cultural development of their own society as they looked to Europe for inspiration, values and validation. On the other hand it seems that peripheral science maintained a long-enduring bias toward the empirical, pragmatic and practical spectrum of scientific activity, an orientation focused on the location-specific data of the immediate environment. What is not apparent is how this schizophrenic focus related to the technological development of the colony.

Models of peripheral science make note of the allegedly narrow focus of colonial science and its contrast with the much more comprehensive and theoretical style of European scientific scholarship. A perusal of the scientific contributions to local societies bears out this view.[1] Yet little analytical attention has been given to the link between science and technology in a peripheral location, and the import this might have for technological dependence. Overall, the literature on Australian science leaves us with two impressions: that science was quite separate from technology; and that scientific and technological dependence or independence ran in parallel, almost as reverse sides of the same coin.

Much research has shown the dependence of Australian science on the metropolitan centre. For this there was often a heavy price to pay, in being kept in a marginal and subordinate position.[2] It has been argued that another price was the acceptance of British ideology and its characteristic disconnection between education, science and industrialisation, which kept an 'elite' science divorced from technical invention, and marginalised both. It is said that while scientific research functions were formed around basically agricultural needs, they nevertheless did little for agricultural technology, minerals exploitation technology or manufacturing industry.[3]

This view is based on an extremely narrow and highly formalised model of science given by the formation of the Council for Scientific and

Industrial Research (CSIR). It ignores prior scientific activity and is somewhat at odds with more recent conceptualisations of science and its
relationship with technology. The differences and boundaries between
science and technology no longer appear so sharp, so amenable to rigid
definition, while their interactions appear more dense, more complex,
more socially conditioned and negotiated, and more equal: no longer can
we accept the simple view that technology is always on the receiving
end of science, simply reducible to the application of prior scientific
knowledge. These understandings are all implicit in the 'seamless webs'
of Thomas Hughes and the 'techno-economic networks' of Michel
Callon,[4] and also the technological writings of economist Nathan
Rosenberg.[5]

Science is not just ideas, theories and knowledge, but also institutions. Science as an intellectual system is not just 'new' knowledge,
but also a whole body of accepted natural relationships, techniques
and methodological approaches. The intellectual system of ideas and
institutions as described by Edward Shils therefore suggests a more
flexible, realistic view of science. Displacing the monolithic model is
a multi-layered structure ranging from the formal, highly organised
institutions such as research institutes, universities, professional and
learned societies—what has been called the superstructure of
science[6]—through to the 'more informal and spontaneous collectivities
through and in which the actual work of productive and reproductive
culture is carried out'.[7]

In Shils' view these latter, ancillary bodies fulfil functions indispensable
for the effectiveness of the major institutions. Mediating between those
institutions and the society at large, they complete the social system of
intellectual life. Through them, the creation and transmission of knowledge has a way into culture and production. Linked up with the major,
formal and obvious institutions of science we can thus see a supporting
infrastructure where government services merge through more informal
structures and activities to general cultural institutions of society. Thus
institutions like observatories, museums and libraries, and publications
like popular periodicals, can support science intellectually by providing
institutional space for scientific roles and training. They can also support
science culturally, providing audience, legitimation, and consequently,
material support from the society as a whole.

Such a view of the scientific enterprise, with its recognition of ancillary
institutions, might be particularly pertinent to a peripheral location such
as late-nineteenth-century Australia. Here the nexus between science and
productive technology might not have been highly visible but it was
certainly multi-dimensional and certainly important for the colony's
capacity to absorb new technology. Indeed several different kinds of links

between science and technology emerge from examination of the two imported technological systems studied in this book.

From science to technological systems

The intellectual content or corpus of knowledge underlying the two technologies was fundamental to their development. In the case of anthrax this involved the knowledge of the anthrax bacillus life cycle, of its causal relation to a particular disease observed in animals, and of the phenomenon of attenuation. The cyanide process evoked the body of knowledge concerning the properties of gold and its solvents, and the relationship of the metals to one another. Along with this body of theoretical knowledge went a repertoire of techniques, in the one case for manipulating bacteria, in the other case for manipulating molecules. These were part of the relevant 'science' and also part of the relevant technology, in the latter case transposed from a laboratory location to more direct confrontation with nature in the field.

Scientists in Australia were not necessarily in touch with the finer points or the newer elements of knowledge embodied in these new technologies. But they had been schooled in the intellectual, methodological and cultural traditions from which they originated. It was a stock of metropolitan knowledge and technique from which they could draw as they cast their eyes to the needs of local production. As bottlenecks arose in technological methods there were several kinds of function which needed to be performed. Local agents of science were a willing resource, as they provided an intellectual bridge to the concepts, perceptions and methods of an advancing metropolitan science and technology.

Who was it that analysed and defined the critical problems which led to the introduction of these two imported technologies? Anthony Willows and Edward Stanley, both men with scientific training, applied their knowledge of bacilli, blood and microscopes to redefine the nature of the periodic disasters taking place in Riverina pastures. In so doing they reoriented the search for solutions. Members and witnesses of Victoria's 1874 Pyrites Board and of subsequent colonial mining commissions applied their knowledge of geology, metallurgy and chemistry to dissect and decipher the failings of local gold extraction.

Similar scientific resources were brought to bear on technological reconnaissance. It was Stanley's bacteriological knowledge which informed overseas inquiries about anthrax vaccines. That was supplemented by the expertise of the Government Analyst, William Hamlet, when the New South Wales Anthrax Board monitored the testing of Pasteur's vaccine. Several other scientific representatives on the Rabbit Commission conducted experiments to screen Pasteur's method of rabbit control.

In Victoria, James Cosmo Newbery brought his Harvard science and his Royal School of Mines metallurgy to his duties as Government Analyst and to his active search for technologies to solve Victoria's goldmining problems. This included screening tests on cyanide for the Royal Commission in 1891. His colleague, Professor Alfred Mica Smith, was a graduate of London and Victoria Universities, and a former research colleague of such scientific luminaries as Bunsen, Playfair, and Angus Smith: he screened the process on behalf of the interests of mining education, the industry, and the belief that educational institutions should lead in matching processes to industry problems. In South Australia Goyder screened the process for a government seeking to stimulate its industry.

New methods had little chance of acceptance while old ways of viewing problems and solutions prevailed. Some means of translating from one mind-set to another was required. Stanley and Hamlet, and then Bruce and Devlin, took prominent roles in persuading stockowners that the decimation of flocks by Cumberland disease was equivalent to the transmission and propagation of the anthrax bacillus. In Victoria, several mining experts translated the decline in gold output to failures of extraction rather than of nature's gold deposits.

If a merging of perceptions between receiving and transferring systems was fundamental to the early stages of diffusion of a new technology, widespread dissemination of information in support of scientific definitions was crucial to maintaining the impetus. Articles describing the 'science' involved in the technologies were part of the transmission of information which facilitated diffusion, and appeared in industry journals and departmental publications.

Scientific method was also a tool, used overtly in the form of trial demonstrations as part of the marketing process. Supporting the Pasteur experts, local officials presented scientific theory and discourse and authenticated the 'scientific proof'. Later, the mystique of laboratory artefacts and manipulations was used to launch the vaccine prepared on Rodd Island. In the cyanide patent disputes, scientific expertise and experiment became an arbiter.

Scientists, both government and private, could also act as consultants, as when users were advised how to adjust the cyanide process to their ores. At other times, they might identify a fault in the technology, as when Pasteur's vaccine was unable to withstand the rigours of Australian conditions.

Meanwhile, others were busy integrating new knowledge and techniques into the local curricula. Edward Rennie, Alfred Mica Smith, Archibald Liversidge, all ensured early entry of the cyanide process into mining education. John Stewart oversaw the introduction of bacteriology and anthrax vaccine technology into the Veterinary Science course at the University of Sydney. A complementary role was performed by those who processed institutional models from overseas. Thus the model of the Pasteur Institute in Paris became translated into the proposal for an Australasian stock institute, while the models of mining schools in London and Freiburg became translated into professional diploma courses in mining.

The scientific community could also exert its influence in other ways. Its authority was used to turn governments against Pasteur's use of disease-based technology, and to introduce legislative controls. Its networks helped bring McGarvie Smith in contact with the Pasteur representatives, and then with Gunn, and its standards and norms provided a measure of Gunn's work. And the formal communication channels could be used just as readily to comment upon government action against the AGRC, as to transmit the results of cyanide research.

In none of the above instances did the link between science and technology appear in the conventionally assumed form of 'scientific' theory providing 'the' principles for development of 'a' technology, or of scientific institutions providing the knowledge base of a technological system. To the extent that that relationship existed, it was all part of the metropolitan development of these technologies. Yet here at the periphery, the local scientific enterprise was acting in diverse ways to smooth the entry of new technology, without creating new knowledge.

Technological research programs and locations

There were therefore many ways in which the local scientific enterprise acted as an intellectual resource for technological systems. Agents of local science were actively engaged in defining problems, scanning overseas technology for solutions and providing supportive activity to the process of assimilation of new technologies into local production systems.

What they were not generally doing was carrying out research aimed at direct solution of technological problems by means of domestic research activity. We look vainly for evidence that the scientific community was engaged in research which took its focus from problems in industry. William Dixon's work in the 1870s stands alone in the formal communications of science as evidence of attempts to initiate a relevant research program prior to the import of overseas technology. Research programs, scientific experiment and theory construction were present, but the trend was to follow the import of technology, rather than precede it.

Whether through lack of interest or lack of confidence, the ready availability of overseas technology seems indeed to have pre-empted, or displaced, local research programs aimed at development of new home-grown technologies to solve an acknowledged industry problem. There was no Louis Pasteur, or even John Macarthur waiting in the wings. There were, however, people ready to seize the initiative a step further along the chain, in two common responses: the initiation of adaptive research programs to resolve the problems arising from the interaction of imported technology with its new environment; and the initiation of comprehensive research programs to reveal more fundamental aspects of the critical problem and its solution.

Goyder, Mulholland and Storer did both, others did one or the other. Donald Clark, director of the Bairnsdale School of Mines, conducted experiments on the solubility of gold in cyanide as a contribution to understanding how the process operated.[8] Jarman and Brereton investigated the use of ammonia to make the process applicable to copper-bearing ores.[9] William Dixon carried out experiments at Sydney Technical College to undermine the selective action theory.[10] The New South Wales Government Metallurgist, James Taylor, and others, tested the hypothesis that gold was not soluble in cyanogen.[11] Among users of the process, we know of the Victorian experiments of George Duncan, John Deeble, W. B. Gray, and White and Simpson. In New South Wales, we know of Henry Watson and Askin Nicholas. In Kalgoorlie there were programs of continuous experimentation in the labs and plants of the large mining companies.

Location might have been significant. Research programs were mainly conducted in the production arena itself or in that part of the infrastructure which lay between industry and the institutions of science. Prominent locations were the laboratories of government analysts, which in South Australia coincided with the School of Mines, in Victoria with the Technological Museum. A similar location is found in the laboratory of the New South Wales Government Metallurgist, at the government metallurgical works. Also within the government infrastructure were the facilities set up on Rodd Island for the Rabbit Commission.

At a slight remove were the laboratories of the Bairnsdale School of Mines and the Sydney Technical College. Then there were the private facilities used by Storer, Mulholland, McGarvie Smith, Gunn, Deeble, Gray, White and Simpson, Nicholas, the Kalgoorlie mines. Closest to the central institutions of science, the chemical laboratories at the University of Sydney housed the work of Jarman and Brereton.

In the reporting of results, the formal channels of local science were only occasionally in evidence.[12] Loir presented two papers to the Royal Society of New South Wales, and McGarvie Smith and Gunn published their paper on anthrax serum in the *Australasian Medical Gazette*, but the results of most of their work appeared in newspapers. Collaborative experiment by Stanley and Loir went mostly into the *NSW Agricultural Gazette*, published by the Department of Agriculture. It was only Loir's connections which took some results back to France to be published in the *Annales* of the Pasteur Institute.

Donald Clark reported to the Royal Society of Victoria, and Goyder published an article through the Royal Society of South Australia, but his other papers went to the journal of the local Institute of Mining Engineers and the annual reports of the South Australian School of Mines. Dixon's work on selective action was presented to the Institute of Mining and Metallurgy in Britain, but was published locally in the *Australian Technical Journal* put out by the Sydney Technical College.

Storer published a cyanide paper in the journal of the Engineers Association of New South Wales, but his work on bromocyanide was, like Mulholland's, published in the pages of the *Australian Mining Standard*. Mulholland did report some further results through the New York periodical, the *Engineering and Mining Journal*, but most of his communications were centred on the Bathurst Science Society, associated with the Bathurst Technical College, and then the Sydney Technical College Mining and Metallurgy Society. Some of these papers also appeared in the local mining journal. That accorded with a general pattern which saw research reports fan out through a spectrum of channels, ranging from those close to the central institutions of local science to the much more peripheral communications of trade journals, and even the daily newspapers.

The science–technology nexus

How, then, should we characterise the science–technology link that formed part of the relations of science and technology at the periphery? Several features have emerged. We have already seen that the scientific enterprise was a vital intellectual resource which facilitated the assessment, entry and assimilation of foreign technology into Australian production. This it did in preference to creating new, local systems of technology.

That facilitating role was particularly relevant when new technology presented a discontinuity with existing systems. Such was the case when imperatives from a young economy compelled the search for radical solutions and the acquisition of advanced technology from the leading edge of the world's technological development. Anthrax vaccination and cyanide extraction were two such cases; they reflected the pressure to move from extensive to more intensive exploitation of resources as the limits of simply moving to new fields and pastures were confronted. But there were many other cases at this stage of colonial development and at a time when organised knowledge was helping to transform the world. Many would be important to the development of a sparsely populated colony. It was, for instance, the need for physical infrastructure which brought railways, telecommunications and urban services such as clean water and electricity.

The technologies that came were on a new plane of sophistication and complexity, and the imperatives which drove their acquisition imposed the challenge of adjustment, but they also opened a door of opportunity. The technological divide would only continue to grow; to take that step across the divide would help put a new economy in touch with the ways and the world of the future. This is the stuff of Perez and Soete's technological windows of opportunity; but it requires people with sufficient understanding to access the required entry knowledge while it is still in the public domain.[13] In Australia, the free flow of compatible cultural resources was a windfall. The local scientific enterprise was schooled in the intellectual, methodological and cultural traditions from which new technological trajectories originated; this not only eased the absorption of advanced technological imports, but enhanced the capacity to learn from them.

It was, however, a role made less visible by its route. Direct relationships between scientific institutions and industry were weak. For the case of the Colonial Sugar Refining Co. (CSR), a contrast has been drawn between the corporate–university relations developing in Germany and America, and the lack of a similar 'symbiotic relationship' in Australia.[14] Yet that company's managerial and chemical leaders intermingled with people from various levels of the scientific enterprise through the meetings of the Royal Society of New South Wales, the Linnean Society of New South Wales and the Australasian Association for the Advancement of Science, all of which heard papers relevant to the sugar industry. It was a Linnean Society scholar, Robert Greig-Smith, who offered his services and laboratory for the microbiological training of CSR chemists.[15] It was an interpersonal connection which led to the enlistment of Professor John Smith's assistance in putting the polariscope to work in monitoring sugar production. And it was the Department of Agriculture which cooperated with the company in work on the control of gumming disease.[16] The colonial way was to interact via the informal networks and the institutional infrastructure which bridged the void between industry and the formal institutions of science.

It was here that many of the experimental programs were carried out and reported. Perhaps this explains the characteristic direction in the way experimental programs developed from the simpler screening and monitoring activities to adaptive programs and on to more comprehensive research. It was much more daunting, much less justifiable, to plunge immediately into the complexities of fundamental mechanisms. Yet simple inquiries and gradual learning could develop a progressively more complex logic.

The people who filled out these patterns also had a common location. Many worked at the juncture between the scientific superstructure, the institutional infrastructure and the productive base. In the persons of government scientists, performing scientific services, the nineteenth-century 'partnership' between government and the private sector can be observed in action.

From technology to science

Colonial science was performing a service for technology, but were there any gains for science? There were, but they were delayed and indirect. They tended to come from the bottom up, in both an institutional and intellectual sense. The need for institutions of science, such as research institutes to study stock diseases, never seemed to be obvious; nor were they likely to emerge from spontaneous support for intellectual endeavour. Activity directed to the needs of technological systems usually appeared first in the infrastructure. Similarly, scientific programs showed no tendency to arise spontaneously from theoretical questions which might bear on technology: as the cyanide process penetrated local mining, no-one in the university delved into the mechanism by which cyanide took gold into solution. Instead, the focus went first to screening and monitoring functions, only later to various levels of research.

So these technological problems did not directly stimulate scientific endeavour, but there was a stimulus nevertheless. Colonial science helped local industry across the technology gap, into processes and products with substantial scientific content. From this the scientific enterprise won its own institutional foothold—not in the major institutions central to scientific production and reproduction, but in the infrastructure which helped support them. Here, experimental programs could be initiated, and scientific personnel could reinforce the relationship between science and technological solutions, carving out a path of future support as they did so.

It is ironic that this stimulus to science arose at a time when colonial science seemed to be making little headway of its own accord. During the 1890s the local scientific enterprise appeared to lose audiences as membership lists of scientific societies typically fell by 20 to 25 per cent. Official support was similarly constrained by slim budgets in a period of economic depression. Yet it was this very depression and its attendant economic disarray which drew government action to promote technologies which offered some hope and relief. Driven by the need to exploit resources more efficiently in the process of recovery, governments gave various kinds of support which worked to enhance technological capability and independence. As it became increasingly apparent that these technologies had some connection with science, the latter gained in credibility. In this sense, the scientific enterprise in Australia developed more from its interaction with technological systems than from its own internal dynamics.

The introduction of the newly emerging science of microbiology serves as a relevant example of this process. It was a coalescence of socio-economic forces in the 1880s and 1890s which hastened infrastructural support for microbiology ahead of its superstructural development. The public health crises in the cities of Melbourne and Sydney in the 1880s were perhaps the most dramatic and impelling of the factors focusing attention on the germ theory of disease. They led in New South Wales to the establishment of the Board of Health, and the drafting of a series of legislative controls affecting several industries. They culminated in the Public Health Act of 1896. In 1897 came a microbiological laboratory and a microbiologist to support the board's new obligations.[17]

In Victoria, the 1889–90 Royal Commission on the Sanitary State of Melbourne was chaired by Melbourne University's Professor of Anatomy and Pathology. Soon after, Thomas Cherry was installed at the university where he operated as an outpost of the Victorian Board of Health. His funding expanded with the services he provided.[18] In a typical sequence, Cherry was appointed in 1905 as Director of Agriculture in Victoria, from which he was appointed in 1911 to the Chair of the newly created School of Agriculture at the university.

Anthrax and the rabbit explosion brought microbiology to prominence in the pastoral industry, while the desire of governments in the 1890s for agricultural diversification brought the appointment of microbiological expertise to foster the dairy industry and agriculture. All this was taking place against a background in which the Senate of Sydney University had rejected the £12,000 bequeathed by William Macleay for a Chair of bacteriology.[19]

Bruno Latour has painted a graphic picture of the way European institutions progressively enrolled, defined and appropriated the science and technology of peripheral regions, reinforcing the power and authority of the 'centre'.[20] While colonial science certainly experienced the force of this process, colonial technological systems seemed able to steer a more independent course. It was notable, therefore, that the path of technological dependence and development was not locked into the course of scientific dependence and development. Indeed, it seems that technological systems may have been running *ahead* of the development of the scientific enterprise, which was still more focused and dependent on a European centre. Technological dependence and scientific dependence did not always move in tune.

Nor was economic dependence an accurate indicator of technological dependence. Inflows of labour were accompanied by inflows of capital, like the flood of British investment into both private and public sectors in the 1880s, and then into mining in the 1890s. They coincided with more informed domestic choice of imported technology, an increase in the range of technological sources and the emergence of clusters of innovative activities. That Australian institutions assembled much of the incoming capital was one important factor; but another was the backing of colonial science, which released colonial technology from the dominating ebb and flow of foreign investment.

Colonial science acted as an intellectual and cultural bridge to the science and technology of the metropolis and thus to the technological systems imported from abroad. In this way it helped assimilate foreign technology into local systems, and helped capture the learning potential inherent in the adoption of sophisticated new technologies. We have seen in a previous chapter that the allocation of suitable resources is a prerequisite of technological learning. In late-nineteenth-century Australia, a readily available colonial science was itself one of the resources allocated to the enhancement of technological capability. There was, in that sense, a free lunch after all.

CHAPTER FIFTEEN

Toward an Australian system

Outcomes from the transfer of foreign technology are never certain. Possibilities range from outright failure of the transfer process, to incompatible and isolated development of foreign and local systems, or successful assimilation into local production. The cases studied here have fallen into the last category—the link was made from the foreign to the local system in a process of integration. Colonial science was an important factor in the assimilation into the receiving industry. So was a public sector whose personnel and structures provided much of the scientific and other support. But how much further into Australian colonial economies were these technological developments able to reach?

Many writers have studied the process by which countries of Western Europe absorbed the industrial technology of Britain and gradually developed their own capacity for innovation and self-generating growth. It appears two factors were fundamental—the development of an integrated economy, and the possession of political sovereignty. The accumulation of changes which fostered the interaction of firms, industries, sectors and regions was vital to opening up the potential of domestic markets, to exploiting external economies, to spreading information and innovation, and also the productivity benefits which in turn added to market demand. Just as important was having the power to determine the conditions of domestic market development. This power depended upon the absence of external control and also on the prior displacement of provincial autonomy and nepotism and the establishment of the rule of law. It was only then that 'exogenous injections of capital, purposive adoption of foreign technology and selective employment of foreign experts . . . introduced dynamic impulses into the economy'.[1]

Technological systems may be viewed at different levels, from the operations of a single firm to the industry-wide networks which penetrate regions or nations. While technology does not necessarily respect political boundaries, there is a body of opinion which accepts that national borders matter. Thomas Hughes and Nathan Rosenberg have described the formation of a characteristic American system of production.[2] Lundvall has suggested several reasons why national borders tend to enclose networks of technological interaction—common government, common legal and institutional constraints and opportunities, common culture, language and education, and also obstructions to the cross-national flow of labour.[3] Common currency is another factor.

The two cases studied here begin to suggest an Australian pattern. Production was dependent on outside technological sources, yet there was an evident capacity to imbibe technology from the frontiers of development—to seek out the new and transfer it in. And, despite the vastness of the land, there was also a capacity for these radically new techniques to spread. Not that the process was immediate. Minds and skills had to be reoriented to new ways of seeing and doing. The technologies themselves had to be shaped to local landscapes. Institutionally, several constraints needed to be removed, including those which bore on the economics of adoption and on the supply of skills. These adjustments took time, some involved struggles for local control of the terms of transfer, but in the process the technological capability of associated actors and industries was raised. There was a capacity for system change.

Australia and its technology have often been derided as derivative, but the very process of adaptation implies a capacity to impress on imported technology features which reflect the imperatives of the Australian environment, manifest in a certain 'technological style'.[4] Some of these features seemed to enhance, others to inhibit the scope for technological independence.

The Australian 'technological style'

The character of domestic science and its relationship with these two new technologies was a positive feature of the Australian 'style'. Of the other features which became apparent as the technologies were absorbed into Australian practice, an obvious one was their typical *small scale*, from the operations of the Pasteur Institute on Rodd Island and the home laboratories of McGarvie Smith and Gunn, to the small cyanide plants put up for £100 or so by individual miners in Charters Towers. Indeed, it was observed that the scale of Australian cyanide operations, compared with the much larger South African norm, inhibited cost reductions.[5]

Distance and transport costs were undoubtedly a factor, denying the scale of centralised processing. In Charters Towers, social structure was also responsible for a situation where something like ninety plants processed the tailings of a single field. The scale of anthrax operations was clearly constrained by the limits of the market, but the limits of finance, and the fact that Gunn had a property to run, prevented even the convergence of the McGarvie Smith and Gunn business into a single operation which might have benefited from economies of scale. The dispersal of the market over a wide area added to the difficulties.

Hand in hand with the lack of size went a certain *lack of functional specialisation*. This has already been noted in relation to research activities which were sandwiched between more pressing duties. In the somewhat *ad hoc* working of the vaccine business and many small cyanide operations there is a sense of a vicious cycle of limited scale, limited turnover, limited finance and limits to specialised divisions of labour. It is notable that it was in the large, well-capitalised mines of Kalgoorlie, with their separate labs and experts, that innovation was most doggedly pursued.

Another distinguishing feature of the way these technologies were assimilated lay in the operation of distinct *colonial sub-systems*. This was not for reasons of any inherent technological imperative, but because colonial parochialism and rivalry were such strong determinants of the Australian milieu, and were manifested in diverging policies and legislation, if not in substance, then certainly in the important details which might affect the operation of technology. It was not only the location of the anthrax belt that constrained the limits of vaccination, but the fact that Alexander Bruce's administrative networks and control extended only to the borders of New South Wales, and his influence only to the government of New South Wales. Victoria's stock administrators, for instance, looked to the import of Parke Davis *anthraxoids* as the way to control anthrax.[6] Even more significantly, the proposal for a national stock institute, modelled on the Pasteur Institute of Paris, foundered on lack of colonial unity. The result was the establishment of *ad hoc*, piecemeal and rudimentary substitutes by New South Wales and Queensland, which subsequently collapsed.

Though the AGRC was an organisation with continent-wide aspirations, it was forced to operate colony by colony. One reason for this was simply distance, that factor which comes up again and again. But over these distances there were clusters of other differing characteristics— different ores, different standard practices into which the process had to be assimilated, different cost structures, different government policies, and different legislation, including patent law. And once again, the different colonial governments, although mostly opposing the AGRC in some way, could not even agree to unite in their opposition.

Along with this characteristic colonial organisation went a strong role for *governments*. That government played a substantial part in the economy of Australia is by now conventional wisdom. What has not been so clearly delineated is how this role was mediated through technology. One way was through the legislation that conditioned the structure and operation of particular industries, and hence the form of the technologies they imported. Less formally, adherence to the 'free trade' line in New South Wales led Henry Parkes' government to reject cash payouts to Pasteur, yet provide many indirect forms of assistance to Loir as he set up his manufacturing laboratory on the government's own property at Rodd Island. To this end, the Department of Mines repeatedly negotiated with the Department of Lands to extend his term of residence from the initial six months to what turned out to be three and a half years. Not least among other forms of assistance were the channels of communication to the market—the pastoral community—through the administrative networks of the stock boards and their sheep districts. Indeed in the working of Bruce, helped by Stanley, one is often reminded of the term 'statesmen in disguise', used by MacLeod to describe those technical experts and administrators who effectively directed the course of events in their field of responsibility.[7] If such 'statesmen' were strongly oriented to science and its potential, they could go a long way to assisting its interaction with technological systems.

The administrative network of mines departments similarly provided a channel for communicating with even the most isolated mining districts on matters of technology. But governments acted to encourage cyanide extraction in many other ways. Sometimes this involved explicit government policy and approval, such as the building of cyanide works, the backing of legal opposition to the cyanide patents, and the purchase of patent rights. Sometimes it occurred through the workings of the department. Goyder's researches began when Inspector Parkes sought to clarify an aspect of the cyanide process for his comparative report on extraction methods. Newbery's successor chose to include cyanide assessments in his laboratory function. Indeed, much of the 'partnership' quality that has been identified in the relationship between the private and public sectors in the working of the Australian economy seems to have operated at this sub-policy level, through the accumulation of many individually small initiatives.

It was also notable that these assimilated technological systems exhibited considerable *local participation* and some measure of *local control*. It was the need for adaptation of the anthrax vaccine which drew Australian participants into its manufacture and eventually put them in control of the domestic market. In cyaniding, the mode of transfer through licensing put the technology into the hands of colonial operators, pushed the responsibility for incremental adaptation onto colonial shoulders, and brought institutional responses which added to colonial control. Here a contrast may be drawn with a colony such as India, where imported technologies remained separated from the majority of local operators. They remained largely cut off from both economic and learning benefits and also from opportunities for institutional change.

Against such a contrast we can set the descriptors which add the finer detail to a picture of Australia's adaptive dependence. Overall, the broad features described above fall into two clusters. The supporting relationship of governments, the resources of colonial science and the local participation in the introduction of imported technology were all related, and extended capability as well as independence of thought and action. Less favourable were the tendencies to smallness of scale, limited functional specialisation and colonial fragmentation. These and the characteristics they reflected could constrain the benefits of imported technology and the scope for stimulating the development of other parts of the economy. These two divergent trends require some further analysis. We will begin with those that favoured technological sovereignty.

Technological sovereignty

Technological sovereignty is not the same as technological independence, or self-reliance. The former implies a freedom to choose, the latter an ability to generate and produce new technology at home. The two may go together, but not necessarily.

Dependency theory cites the concentrated supply of inputs from a single source as a major indicator of dependency. Here we found that Australia looked to a number of different countries for its technology, institutional models and equipment. In the anthrax case, France supplied the technology, some of the science and institutional models, and also the capital and people required to establish vaccine production in Australia. Germany supplied another perspective on institutions and other aspects of the science, its techniques and equipment. Gunn's incubators, for instance, were of a German make and described by a government scientific officer as 'not the type usually used in British laboratories'. McGarvie Smith had incubators made up in Sydney but took advantage of Germany's instrumental skills for the manufacture of his own design of syringe. In the cyanide case, Britain supplied the technology, and the capital and experts necessary for its introduction, but other experts came from Germany, the United States and South Africa, as did much information. More generally, Germany provided the most highly esteemed institutional model for mining education, while America provided the model for the professional mining organisation.

Australia's need to import all these factors was a sign of its technological dependence. However, this increasing diversity in sourcing had implications for Australia's technological sovereignty. It is hardly surprising that a nation formed as a colonial offshoot of Britain would receive a large proportion of its people, culture, institutions, technology, capital and ideas from Britain. The common bonds of culture and language, added to all the commercial, technical and personal connections between the two countries, naturally made Britain the first port of call. The crucial issue of sovereignty turns on whether it was a case of independently and freely *seeking out* what was needed or whether the element of choice was eliminated by political controls, by economic controls, or the ideological control of Britain's cultural hegemony.

The Australian colonies were certainly loyal to all things British. Yet practical priorities could intercede. Anthrax vaccination was *chosen* after an international round of inquiries which deliberately by-passed the Colonial Office. New South Wales' own Agent-General in Britain helped the flow of communication, but most subsequent links were direct from Australia to Paris. Britain served largely as a source of English-language bacteriological texts. The cyanide process had all the advantages of colonial connections but had to prove itself to Australian mining against the competition of other overseas technology. It also found itself in competition with the local Newbery–Vautin method of chlorination, which, it should be remembered, maintained its hold at the all-important site of Mount Morgan. The cyanide patentees also had to run the gamut of colonial patent laws and government responses to local lobbying in order to gain access to Australian profits. In the end, the holders of the Australian patents made only a meagre profit, forced as they were to fight all the way over the terms on which the process would be used. The reality hardly fits the picture of the all-powerful and dominating foreign corporation portrayed during the royalty campaign. Instead, the AGRC faced an alignment of diverse interests riding on the back of a rising nationalism, and governments grasping at opportunities for economic recovery. Those governments demonstrated considerable effective political sovereignty in their response to the attendant technological challenges and conflicts.

Though British technology did have obvious colonial advantages over the products and processes with which it competed on the international technology shelf, its choice for Australian systems was neither inevitable nor a reflection of imperial coercion or ideological tunnel vision. Instead, through the support of colonial administrations and colonial science, there was a developing internal capacity to analyse production problems and seek out specific solutions in an informed way. Another element of technological sovereignty came from the capacity to monitor the performance of imported technology, to identify problems in its interaction with a new environment, and to solve these problems by adapting the technology, often generating innovations which were more than just marginal adjustments. On a cumulative basis, the inputs of technology and people were combining to build local systems with an increasing capacity to select, adopt and adapt. As they did this, they gained increasing capacity to control and generate change from within. The question is, can we generalise from two cases?

The bigger picture

The sovereign selection of technologies from an international technology shelf is visible in several areas. In brewing, British practice and technology had been established from early times. By the early 1880s the local industry had already begun breaking away from British methods and developing techniques better suited to the Australian climate and raw materials. The use of cane sugar, an advantage because of the high nitrogen levels of local worts, was particularly indicative since this had been banned in British brewing since 1847. Just as significantly, it was French science and Danish technology which provided the basis for solving one of the major problems of colonial brewing—that of unwanted secondary and tertiary fermentations arising from the yeast. Imported in 'blueprint form' from Copenhagen, E. C. Hansen's techniques allowed local brewery scientists to solve their immediate production problem and also triggered local research programs. These led to the discovery of an Australian wild yeast, to the demonstration that Hansen's methods were applicable for top fermentation, and to the production of Australian Yeast No. 2, the first pure yeast used commercially in Australia. In fact, Hansen's work found application much earlier in Australia than in Britain. By the early 1900s Melbourne chemists had produced three new specially designed yeasts: Melbourne No. 1 could absorb 50–60 per cent of cane sugar without weakening or allowing secondary fermentation; Melbourne No. 5 met the needs of the Abbotsford Brewery; and Melbourne No. 6 was designed for lager brewing.[8]

   In sugar production, a shift from simply refining imported sugar to milling local sugar-cane drew on the milling technology of the West Indies, Britain and France. An eye was also kept on the application of organised science to the beet-sugar industries of France and Germany. Though not himself a scientist, E. W. Knox kept himself scientifically informed, partly through membership of scientific societies. He recruited the first chemists for the Colonial Sugar Refining Co. (CSR) from Scotland, but required familiarity with French practice. He then decided that a 'first rate German chemist' would be ideal, and recruited two Germans, one with a Ph.D. Far·from inhibiting internal innovation, the imported milling technology set off a train of experiment and organisational change, including internal training of local chemists. One of the innovations that resulted was a chemical method of process control that put CSR at the frontier of sugar technology. When the company moved into cane growing, it boosted the demand for crushing machinery, evaporating pans and centrifugal equipment from Mort and Co. and P. N. Russell and Co. in Sydney.[9]

In the dairy industry, the Scandinavian cream separator and the American Babcock fat tester combined with the Australian refrigerator to create the technological basis for the major transformation manifest in the rise of the factory system. Experts installed by governments eager to advance the industry added to the momentum. After a trip to Europe in the early 1890s, Victoria's David Wilson urged the adoption of the French technique of pasteurisation.[10] The British dairy expert recruited for New South Wales insisted that the 'technique of Australian dairying requires to be more guided by bacteriological research than the dairy work of any other country with which Australia competes'.[11] He encouraged its application to product improvement and development throughout the industry, and carried out his own investigations in his government laboratory. Soon he was supplying the industry with starter cultures.[12] Meanwhile, the Sydney engineering firm of Waugh and Josephson was using the local agency for Danish separators to build a business and reputation as a custom supplier of dairy machinery and equipment.

A growing wheat industry spawned a cluster of innovative activities and linkages based around a series of imported techniques. The well-known plant-breeding of William Farrer was supported by the skills of Frederick Guthrie, the chief chemist of the New South Wales Department of Agriculture. Trained in London and Germany, he drew on Danish methods of protein determination and British methods of protein fractionation to develop acclaimed procedures for determining the milling and baking qualities of wheat and flour. Meanwhile, flour millers were importing the latest technology, such as roller mills, but local engineer Robert Bodrington became so adept at roller milling technology that he developed a number of highly praised patent machines and was able to compete with the famous Ganz and Co. of Hungary, Cornelius International Roller Mills of the USA and Henry Simon of Manchester. In harvesting, Hugh McKay's adaptation of mechanical harvester designs to Australian climate and terrain resulted in machinery which could defeat foreign competitors both at home and abroad. In biscuit making, where automatic machinery was being introduced by the 1870s, T. B. Guest and Co. won favour with their pattern for the machine for making 'Currant Luncheons', copied by English manufacturers of biscuit machinery.[13]

Superphosphate technology was carried from Rothamstead to Victoria in the 1870s by James Cumming, a Scottish veterinarian who established a modern complex producing concentrated sulphuric, nitric and sulphurous acids, ammonium nitrate and phosphate fertiliser. His son learned chemistry in Melbourne's Public Library and by working on the construction of the acid plant. Its success and efficiency was evident from the firm's several industrial awards and its export of calcium phosphate to Mauritius and New Zealand. But widespread use of superphosphate from the 1890s awaited the development of application methods suited to local conditions and the scientific demonstrations offered by local agricultural colleges. By 1901 there were two factories in South Australia. By 1930 eighteen Australian superphosphate plants had a capacity of more than 1 million tonnes per year.[14]

It may be argued that these examples all relate to the processing of primary products and only confirm the narrowness of Australia's industrial base, yet there was also evidence of more broadly-based innovation. Though Australian engineering had been an extension of British engineering, by the 1890s the novel bridge designs of Percy Allan, including the use of timber trusses, symbolised a generally more independent orientation to local conditions and a desire to draw inspiration from many countries.[15] The opening of the electrically operated span of the Pyrmont Bridge in 1902 was his crowning achievement. Electrical technology showed similar trends. Materials, designs and knowledge transferred from United States, British and German sources were combined with local inventiveness and the support of the scientific community to form the Victorian company that was to pioneer electrical technology in Melbourne. It proved itself technically superior to overseas competitors.[16] In telecommunications, early electromagnetic experiments by William Bragg and Richard Threlfall helped sensitise the founders of the federal constitution to the implications of wireless telegraphy and the need for a national legislative framework. By 1914, the nineteen wireless stations around the Australian and New Guinea coast were mainly using the telegraphy technology of the Australian engineer, John Balsillie.[17]

Thomas Baker, a young Victorian chemist, saw the opportunity in American George Eastman's production of dry photographic plates, which displaced the cumbersome wet-plate process. In 1887 Baker joined with accountant John Rouse, who set up a network of outlets while Baker concentrated on process improvements in the laboratory: results included Austral dry plates, an improvement on the originals, and the Australian-made Simplex gaslight paper, an improvement on the American Aristo paper. In other chemical developments, the new

explosive material cordite was found vulnerable to the high temperatures and humidity of Australia, leading to the adaptation of the
manufacturing process and establishment of local manufacture in
1908.[18]

It was the desire to exploit rich iron-ore deposits in South Australia
which led BHP into heavy industry with the establishment of iron and
steel works in 1915. The company took advantage of good technological
timing, bringing the modern open-hearth furnaces which were replacing
Bessemer converters overseas and also the new basic furnace practices
which were replacing acid slag practice. This made it possible to produce
better low-phosphorus steels. Soon makers of steel products clustered
around BHP. But steel-related products were not new in Australian industry. Thomas McPherson's company began nut and bolt production with
plant from America's Acme Machinery Co. in 1900 and experts imported
to commission it. Later innovation brought a cold forging process and a
substantial improvement in quality, as well as an expanding range of
products, from pumps to machine tools and grinding equipment.[19]

A leap forward from dependence on the blacksmith's forge to the
radical new technology of welding metals began when Russell Grimwade, a Melbourne science graduate, journeyed to England and saw
Dewar's experiments with the storage of liquids at low temperature and
Hampson's apparatus for liquefying air. The liquid air plant ordered in
1909 for the family firm of Felton, Grimwade and Co. was a German
model, installed by a German-trained Hungarian. In the same year, the
Scottish-trained engineer, William Flyvie, joined Grimwade on the platform of the Institution of Engineers when he demonstrated oxy-acetylene
plant brought back from the Franco-British Exhibition in Paris. Flyvie was
the channel by which the British Oxygen Co. brought German, French
and British technology to Australia through its Sydney subsidiary Commonwealth Oxygen Co. Ltd. Swedish plant for manufacturing acetylene
came to the Melbourne firm of Gardner Waerne and Co. Together, these
developments were the beginning of a new industry which took Australian welding technology to the world frontier.[20]

As other nations industrialised and offered more appropriate technological options, Australian industries were exerting their sovereign right
to choose, a trend also evident in the declining proportion of imports
from Britain.[21] In so many cases, the imported technology then triggered
local technological activity, innovation and learning, as it was adapted to
the local environment with the help of the informal support of colonial
science. This in turn added to domestic technological capability and reinforced technological sovereignty. Such factors gave Australia in 1904 6.4
patent applications per 10,000 of population, in comparison with Britain's 7.0 and the USA's 6.6: by 1916 the respective figures were 7.1, 4.3

and 6.7.[22] The trouble lay in converting this inventive activity into commercial successes in the market.

The problem of size, density and demand

The first cluster of features of the Australian 'technological style' added to the scope for enhancing technological capability and sovereignty. The second cluster of features, however, suggested opposite pressures. Small scale, limited functional specialisation and colonial fragmentation were all related and affected that vital driving force—market demand. It was a circular effect. If demand was small, the scale of production was small and its costs accordingly high, making it more difficult to compete with imports, making it less attractive as an investment. The colonial segregation which fragmented the country's infrastructure, and split the country's market, only added to the effects of a small population dispersed over a vast land. As the manager of Ballarat's Phoenix Foundry revealed to a royal commission in 1882, he 'could manufacture anything at all you like, but if you put yourself in the way of manufacturing, say lathes, or boring machines, or planing machines, or anything of that kind, it must pay you to do that and nothing else'.[23] The problem for his company was the irregularity of orders.

A typical example demonstrates the spiral that inhibited colonial enterprise. By the early 1860s there was considerable pressure within New South Wales for government contracts to be placed with local manufacturers, especially the large sums to be expended on locomotives and rolling stock for railways. But the fundamental problem confronting anyone contemplating locomotive construction, as distinct from assembly, was the necessity for some assurance of continuity of orders to justify the risk of investment in plant and equipment. From the government end, the difficulty in placing contracts locally was that colonial prices were bound to be higher than those obtained from Britain, where large establishments had the benefits of extensive division of labour, and of labour costs only one-third those in the colony. No doubt British firms also had the flexibility to tender low in some markets and cross-subsidise from others. In addition, the raw material had to be imported from Britain.

Even so, the two firms of Vale and Lacy, and Mort and Co. were given orders for locomotives although their tenders were 22 per cent higher than overseas quotations. They had tendered on the understanding that at least thirty-five engines would be required over a five-year period, but they built only a total of twenty-two. They were forced to import and assemble another twenty-four engines in order to meet delivery deadlines. Irregular timing of orders was due to government budgetary problems and uncertainties about railway policies which led to sudden emergency orders. This placed a great strain on colonial manufacturers who were unable instantly to summon skilled labour, and who were plagued by delays in deliveries of components from Britain. Rising local wages and a sharp rise in the cost of British components added to local costs. When the contract lapsed in mid-1874, even such an enthusiast as Thomas Mort said it would not be

> wise under existing circumstances . . . to bring out the machinery required for the manufacture of the wheels, steel axles and tyres. Had we to erect the costly plant to make these for our own comparatively small wants, it would involve an outlay, the interest on which would tell against the industry very seriously.[24]

Governments were caught in the delicate balance required between supporting local industry and laying themselves open to the charge of squandering public money, and the circular trail of events kept repeating. In the 1880s a series of negotiations on tenders eventually awarded the contracts to local firms after they had reduced their tender prices from 60 per cent above British quotations, to 50 per cent and then, finally, 30 per cent above. Yet within a year both the Sydney firms had relinquished their contracts without delivering a single locomotive. Such were the frustrations faced by colonial manufacturers and their customers.

Yet the picture was not all bleak. Even imported technologies could help generate change if they set up a connection between the needs of one industry and the stimulus to the growth of another. In the case of anthrax vaccination, McGarvie Smith and Gunn's response to a problem in the pastoral industry provided a stimulus into manufacturing, with a business based on the newly emerging science of bacteriology. Their success attracted others, and by 1918 there were at least three firms in Sydney supplying anthrax vaccine. Though this supply industry was small in scale, it led on to vaccine businesses which became exporters—the McGarvie Smith Institute and Arthur Webster Pty Ltd. Through diversifying into a range of vaccine technologies, the latter became a multi-million-dollar business marketing a whole range of vaccine products at the leading edge of technology. The Commonwealth Serum Laboratories, which developed as a public competitor, has recently been privatised, bringing a purchase price in the vicinity of a billion dollars.

In the case of cyanide, several engineering firms who were already supplying mining equipment extended their product range by manufacturing cyanide plant. Sometimes they incorporated their own innovations. Though not highly complex, this work provided a boost to firms buffeted by depression. Victoria's Jacques Bros was one company saved from oblivion in the 1890s when it switched its focus from the decimated construction industry to the booming mining industry. With the increasingly sophisticated skills and machinery thus developed, it was able to shift focus once again in the twentieth century, moving to the manufacture of heavy machine tools, excavators, power shovels and motor cranes when mining activity declined.[25] The embryonic chemical industry was not so fortunate. Mining laboratories might use local acids and alkalis, but cyanide users were tied by contract to Cassel supply for the life of the patents. By the time these terminated, supply lines back to Europe were well established. It was indeed as a supplier of cyanide that the Cassel Company made its own fortune and ended up as part of the ICI network.

More generally, however, colonial integration was improving, if not surging ahead. Colonial systems were interacting more. Intercolonial stock conferences arose from the need for unity to protect Australia from the introduction of foreign disease, and found an increasing number of issues to deliberate. Sometimes there was co-operation, often there was not. The conference agreed to support the statement of principle that Pasteur's anthrax vaccine had been proved effective, but could not agree to fund the purchase of its method of manufacture. Conference members likewise agreed in principle on the desirability of a national stock institute, but failed to follow up with action in their own domain. Goldmining colonies also observed neighbouring policies, yet did little to advance intercolonial collaboration, even against a common foe such as the AGRC. By contrast, the AGRC was an Australia-wide organisation, seeking a national market, using any cross-colonial connections it could find, and moving people across the length and breadth of the continent.

Other forces for unity emerged with the national industry journals and national professional associations and ran in parallel with the end-of-century move to Federation. Here the transfer of communications to federal control was an important advance, as was the establishment of a federal patents administration. On the other hand, the states remained the major developmental agencies, retaining most of the relevant responsibilities. For most production, colonial boundaries remained effectively intact, even though there were now mechanisms, such as fiscal tools, for urging greater collaboration.

Technology transfer into so-called 'regions of recent European settlement'

Australia was, of course, a land assumed to be effectively 'empty' when British culture was brought to its shores. It has in the past, therefore, been labelled as a 'land of recent settlement', a term applied to a group of countries which includes the United States, Canada and New Zealand.[26] These countries all possessed large open grasslands and vast unexploited natural resources, and Europeans made permanent and populous settlements which overwhelmed any prior inhabitants by their sheer weight of numbers as well as force. It is now clear that the label is inappropriate; one suggested alternative is 'land of recent invasion',[27] and the point it makes is a valid one. Even more important for our story is the point that the invading culture was so victorious that those countries soon wore a European face; it was almost as though they had been previously uninhabited. In this they differed from more densely populated colonies, and that difference was significant for the long-term political and economic outcome. The difference was also significant for the process and effect of transferring foreign technology, for in those 'empty' lands there were both advantages and handicaps for the conversion of transferred technology to autonomous growth.

Many nations, especially those of Western Europe, used technology transfer to catch up with industrial leaders. As pointed out in the first chapter, though, the transfer of technology alone could achieve little in bringing development to any country. What helped those which successfully followed Britain into industrialisation was the adoption of deliberate state action to avoid being swamped by Britain's superior technological competence. With the benefit of political control, technological dependence could be transformed into creativity, self-generating growth and all-round autonomy.[28] At the other extreme, many of those countries colonised by European powers had technology transfer imposed upon them as part of the package of imperialism, with the loss of political and/or economic sovereignty. Their ancient established technological systems were rendered irrelevant or incompetent, and overwhelmed by those of the invading culture.

Australia fits a different category. Here in this allegedly 'empty' land, the rationale for technology transfer was not simply for catching up but for growing. It was not simply for replacing old, or 'inferior' or 'primitive' technology, but for building technological mass. The technological dependence of the British settlement was therefore more inevitable, more pressing and more pervasive than elsewhere. This is where the advantages were important. The advantages came with the free flow of European culture, institutions and labour. Coming from the leading industrial centres of the world, they supported, nourished and assimilated the advanced technology brought from the same source. Furthermore, the 'metropolitan' life and technology which expanded into vast open spaces met with relatively little resistance from a dispersed and unobtrusive existing culture and technology. Though interpretations of the contact are now being revised,[29] the contrast with the crowded clash of cultures in places like India and China is extreme.

Yet the very 'emptiness' had its problems. Australia was not merely small in population; that population was scattered over a large area, even though much of it clung to the eastern seaboard. There is now a body of research which shows the importance of geographical concentration—of markets, of related production and innovation activities, and hence of the interaction between them—in creating the competitive advantages which stimulate dynamic growth.[30] As Japan went seeking the benefits of foreign technology, it could feed its technological imports into dense agricultural and handicraft sectors. Australia could never hope to become self-sustaining until it reached some critical size and density. Therefore, the extent of its growing political sovereignty was important, but perhaps even more crucial was the capacity to develop the interaction between firms, industries, sectors and colonies which would co-ordinate infrastructures, yield an integrated economy, maximise the market and develop the conditions for a broadly-based self-generating growth.

Australia could absorb foreign technology into the domestic economy: the greater difficulty lay in sparking dynamic interaction from the receiving sector to others. This was a constraint rather than a blockage, but the challenge lay in reaching that threshold of geographically concentrated market demand on which commercial viability would depend. Yet even when the demand for producer goods was expanding, the introduction of machinery from diverse sources had the effect of fragmenting the market for materials and minor components that were technically within the competence of local firms. This applied not only to the different sizes and threads of bolts and screws, but to the kinds of glass containers suited to different sorts of bottling plant, to the pegs and rivets needed for different kinds of boot-making appliances, and even to the nails that could be used with door and sash-making machinery. For this reason, some firms added on their own workshops to cater for their own specific needs.[31] This response reduced imports, but undermined potential economies of scale from specialised suppliers.

Internal colonial policies could also be a constraint on the development of an integrated and balanced industrial economy. As Slater has shown for the supply of electricity to Melbourne, the priorities of rival factions in the city council kept electricity off the streets, while opposing factions within Victoria's parliament undermined the Australian company and the substantial skills it had developed. Local politics, rather than technical factors, drove the assessment of rival technologies, and laid the grounds for the eventual victory of the foreign product.[32] More generally, Linge has shown that the implementation of colonial policies of local placement of government contracts was in practice often subject to the political and other exigencies of the moment, thus maintaining an environment of uncertainty for colonial investment.[33] Another significant failure in local policy was manifest in the inflexibility of customs schedules, which seldom distinguished between goods like oils and paints intended for final consumption and seemingly similar materials required by manufacturers.

Intercolonial rivalry also had its effects: the desire to divert traffic from neighbouring colonies was part of the background which resulted in disconnected colonial railway systems with differing gauge sizes, uncoordinated schedules and fluctuating freight rates. Within colonies, rivalries between towns and regions influenced government decisions on such important investments as branch railways, bridges, water schemes and harbour works. Local policy and politics, at various levels, could exacerbate the costs of geographical distance for colonial manufacturers, break down linkages and density that were tentatively being developed, and undercut the potential for specialisation in a capital goods industry.

Discarding certain dependency argument myths

We must therefore cast aside certain myths that have lingered as the result of applying dependency theory and models to the Australian experience. The most broad-ranging is the myth that Australia's position of dependence was somehow analogous to that of colonised countries with large, established indigenous populations and cultures, those which are now classified as 'underdeveloped', 'less developed' or 'developing'. With that came the notion that British colonisation created a structurally dependent economy which, because of its dominance by British and pastoral interests, was always partial and unbalanced in its development and destined for continuing imperialist exploitation. The companion of that idea was the view that Australia was technologically primitive and backward, a passive victim of the needs and interests of Britain. With it went the claim that Australian science in the late nineteenth century was divorced from local productive technology.

Dependency theory has done much to reveal the structural imbalances and inequalities that resulted from various phases of imperialism, but its application to Australia leaves important features of the Australian experience out of the picture. One of the most significant is the chronic problem of size, density and demand. For, though Australia was built on migrant populations, the influx of Europeans to Australia could never compare with the millions who made the much shorter journey across the Atlantic to North America. Nor could its resource, production and population base develop the transport systems which helped create a huge and integrated national market.

Even so, Australia's nineteenth-century profile of development was in fact similar to that of an earlier America—predominantly rural, exploiting natural resources, limited mostly to consumer-oriented manufacturing, and growing directly from immigration, foreign capital imports and associated transfers of European technologies. It was closer to the American model than it ever was to those of India or Africa, where all kinds of subordination to a foreign power overlapped. Australia's transplanted British society was instead accumulating increasingly meaningful political sovereignty. Moreover, its scientific, technological and economic dependence, while related, were not locked together.

That is not to say that a young immature system was not *vulnerable* to the sheer economic weight of British dominance of the international economy. But knowledge is power, and colonial science was a factor which could reduce vulnerability and turn the transfer of foreign technology to advantage. Supported by the resources of local science, Australia showed itself technologically robust, and an active and informed scanner of the international technology shelf, seeking advanced technologies to upgrade its production and infrastructure.

Far from being passive victims of British manipulation and exploitation, local producers were prepared to engage in intense struggles for control of the technology they imported, often with considerable success. And while the export sectors, which favoured British interests, did undoubtedly still dominate the Australian economy in the late nineteenth century, this imbalance was being modified as other parts of the economy were developing. Part of that increasing diversity arose because technological imports into the export sectors could and did lead to local innovations which in turn led to developments in local manufacturing. We should not be misled by the fact that they were small in scale. Nor should we misinterpret the hindrances they suffered. For the latter came as much from local impediments as from the perversity of British policy and capital or the complicity of pastoral interests.

The Antipodean reality has always been more than the plaything of imperialist policy and ideology. Our stories on anthrax and cyanide showed a complex interplay between foreign and imperial desires, intentions and influences, and the local interests and imperatives they confronted. In this, the 'periphery' has been as important as the 'metropolis', and far less marginalised than many countries that lay geographically much closer to the European 'centre'.[34] It has indeed 'developed as a province *within* the metropolitan economy'.[35] Australia would, in a sense, create its own wheel of fortune as it sent its technological envoys into a multipolar world. That it was able to enrich itself on what they brought back was in part a reflection of that cultural bridge that had joined the developing technology of the periphery to the advanced science and technology of developing and multiple metropolises.

Notes

1 DEPENDENCY AT THE PERIPHERY

1. For a collection of essays which reflect a range of views on this dilemma, see Martin Fransman and Kenneth King (eds), *Technological Capability in the Third World* (London, 1984).
2. C. A. Mulholland to Editor, *AMS*, 3 August 1895, 441.
3. A. Barnard, 'A Century and a Half of Wool Marketing', in A. Barnard (ed.), *The Simple Fleece* (Melbourne, 1962), 476.
4. W. A. Sinclair, *The Process of Economic Development in Australia* (Melbourne, 1976), 83; Barrie Dyster and David Meredith, *Australia in the International Economy* (Melbourne, 1990), 27.
5. Dyster and Meredith, *Australia*, 41.
6. *Ibid.*, 40-9.
7. Paul Grant, 'Technological Sovereignty: Forgotten Factor in the Hi-Tech Razzamatazz', *Prometheus*, 1 (1983), 239-70.
8. Brian Fitzpatrick, *The British Empire in Australia, 1834-1939*, 2nd edn (Melbourne, 1969), 299.
9. N. G. Butlin, *Australian Domestic Product, Investment and Foreign Borrowing, 1861-1938/9* (Cambridge, 1962); *Investment in Australian Economic Development, 1861-1900* (London, 1964); 'The Shape of the Australian Economy, 1861-1900', *Economic Record*, 34 (1958), 10-29.
10. N. G. Butlin, *Bicentennial Perspective of Australian Economic Growth*, Canberra, Economic History Society of Australia and New Zealand, Inaugural public lecture, 1986.
11. Encapsulated in W. W. Rostow, *The Stages of Economic Growth* (Cambridge, 1960).
12. A well-known version is found in André Gunder Frank, *Capitalism and Underdevelopment in Latin America* (New York, 1967).
13. Dieter Sengaas, *The European Experience* (Leamington Spa/Dover, New Hampshire, 1985), 156.
14. David Clark, 'Australia: Victim or Partner of British Imperialism', and E. L. Wheelwright, 'Introduction', in E. L. Wheelwright and Ken Buckley (eds), *Essays in the Political Economy of Australian Capitalism*, vol. 1 (Sydney 1975); David Clark, 'Unequal Exchange and Australian Economic Development: An Exploratory Investigation', *ibid.*, vol. 3 (1978); Peter Cochrane, *Industrialization and Dependence* (Brisbane, 1980).
15. George Basalla, 'The Spread of Western Science', *Science*, 156 (1967), 611-22.

16. Ann Mozley Moyal, *Scientists in Nineteenth Century Australia* (Melbourne, 1976); M. E. Hoare, 'Light on Our Past: Australian Science in Retrospect', *Search*, 6 (1975), 285-90, and 'Science and Scientific Associations in Eastern Australia, 1820-1890', unpublished Ph.D. dissertation, Australian National University, 1974.

17. Roy MacLeod, 'On Visiting the Moving Metropolis: Reflections on the Architecture of Imperial Science', *Historical Records of Australian Science*, 5 (3) (1982), 1-6.

18. Ian Inkster, 'Scientific Enterprise and the Colonial "Model": Observations on Australian Experience in Historical Context', *SSS*, 15 (1985), 677-704.

19. Donald Fleming, 'Science in Australia, Canada, and the United States: Some Comparative Remarks', *Proceedings of the 10th International Congress on History of Sciences* (Ithaca, NY, 1964), 179-96.

20. A leading proponent was E. L. Wheelwright, as, for instance, in 'Development and Dependence: the Australian Problem', *Australian Quarterly* (1971), 22-39. As a forerunner, see Brian Fitzpatrick and E. L. Wheelwright, *The Highest Bidder* (Melbourne, 1965). The Jackson Committee was also concerned in its *Policies for Development of Manufacturing Industry*, vol. 1 (Canberra, 1975), 98-9.

21. M. L. Alexander, 'Australia in the capitalist world economy', in Brian Head (ed.), *State and Economy in Australia* (Melbourne, 1983), 55-78.

22. Ted Wheelwright and Greg Crough, 'The Political Economy of Technology', in Roy MacLeod, *The Commonwealth of Science* (Melbourne, 1988), 334.

23. For instance, Stephen Hill, *The Tragedy of Technology* (London, 1988). Less specifically focused on technology but advancing the same kind of perspective are Cochrane, *Industrialization*; Clark, 'Australia: Victim'.

24. Brian Johns, 'Technology as a Resource. Inventing Less, Importing More?', in Peter Hastings and Andrew Farran (eds), *Australia's Resources Future* (Melbourne, 1978), 162-84; Tuvia Blumenthal, 'A Note on the Relationship Between Domestic Research and Development and Imports of Technology', *Economic Development and Cultural Change*, 27 (1979), 303-6; D. T. Brash, *American Investment in Australian Industry* (Canberra, 1966). On the industry-specific nature of this effect see Peter Stubbs, *Innovation and Research* (Melbourne, 1968).

25. K. G. Gannicott, 'Research and Development Incentives' in Myers Committee of Inquiry, *Technological Change in Australia*, 4 vols (Canberra, 1980), vol. 4, 287-314.

26. *Ibid.* See also P. J. Morris, 'Australia's Dependence on Imported Technology—Some Issues for Discussion', *Prometheus*, 1 (1983), 144-59. For a collection of papers analysing related issues see T. G. Parry (ed.), *Australian Industry Policy* (Melbourne, 1982).

27. David Landes, *The Unbound Prometheus* (Cambridge, 1969), 1.

28. Sidney Pollard, *Peaceful Conquest: The Industrialization of Europe, 1760-1970* (Oxford, 1981), 142.

29. e.g. E. Denison, 'United States Economic Growth', *Journal of Business*, April 1962, 109-21.

30. David Jeremy, *Transatlantic Industrial Revolution: The Diffusion of Textile Technologies Between Britain and America, 1790-1830s* (Oxford, 1981), 76.

31. F. Crouzet, 'Western Europe and Great Britain: "Catching Up" in the First Half of the Nineteenth Century', in A. J. Youngson (ed.), *Economic Development in the Long Run* (New York, 1972), 98-125.

32. An influential work in that discourse was Barry Barnes, *Scientific Knowledge and Sociological Theory* (London, 1974).

33. Thomas Hughes, 'The Seamless Web: Technology, Science, et cetera, et cetera', in Brian Elliott (ed.), *Technology and Social Process* (London, 1989), 106-19; D. MacKenzie and J. Wacjman (eds), *The Social Shaping of Technology* (Milton Keynes, 1985). The trends of intellectual development are evident in the journals *Social Studies of Science* and *Technology and Culture*.

34. Barry Barnes, 'The Science-Technology Relationship: A Model and a Query', *SSS*, 12 (1982), 166-72.

35. A volume dedicated to this approach is W. E. Bikjer, T. P. Hughes, T. Pinch (eds), *The Social Construction of Technological Systems* (London, 1987).

36. Thomas Hughes, 'The Evolution of Large Technological Systems' in *ibid.*, 51-82, and 'The Order of the Technological World', in A. R. Hall and Norman Smith (eds), *History of Technology*, 5 (1980), 1-16; Nathan Rosenberg, *Inside the Black Box* (Cambridge, 1982), 59; Wesley Schrum, *Organised Technology: Networks and Innovations in Technical Systems* (West Lafayette, 1985), 15, and 'Scientific Specialties and Technical Systems', *SSS*, 14 (1984), 63-90.

37. Thomas Hughes, *Networks of Power* (Baltimore, 1983).

38. Thomas Hughes, 'The Evolution of Technological Systems', in Bijker *et al.*, *Technological Systems*.

39. Robert Evenson and Hans Binswanger, 'Technology Transfer and Research Resource Allocation', in H. P. Binswanger and Vernon W. Ruttan (eds), *Induced Innovation. Technology, Institutions and Development* (Baltimore, 1978), 164-211.

40. Thomas Kuhn, *The Structure of Scientific Revolutions* (Chicago, 1962); Joseph Schumpeter, *Capitalism, Socialism and Democracy* (New York, 1942).

2 CROSS-CURRENTS OF CHANGE

1. Richard Twopeny, *Town Life in Australia* (London, 1883), Penguin 1973 edition, 148.

2. Barrie Dyster and David Meredith, *Australia in the International Economy* (Melbourne, 1990), 22, 34.

3. N. G. Butlin, *Investment in Australian Economic Development 1861-1900* (Cambridge, 1964), 22, 201-10.

4. D. N. Jeans, 'The Impress of Central Authority upon the Landscape', in J. M. Powell and M. Williams (eds), *Australian Space, Australian Time: Geographical Perspectives* (Oxford, 1981), 1-17; M. Williams, 'More and Small is Better', *ibid.*, 68-9; Alan Gilbert, 'The State and Nature in Australia', *Australian Cultural History*, 1, 9-28.

5. Colonial Secretary, Earl Grey, quoted by I. D. McNaughtan, 'Colonial Liberalism, 1851-92', in Gordon Greenwood (ed.), *Australia: A Social and Political History* (Sydney, 1977), 99-100.

6. T. H. Irving, '1850-70', in Frank Crowley (ed.), *A New History of Australia* (Melbourne, 1974), 124-64.

7. On this process of erosion see McNaughtan, 'Colonial Liberalism', 131-4.

8. W. J. Hudson and M. P. Sharp, *Australian Independence: Colony to Reluctant Kingdom* (Melbourne, 1988), 39-40.

9. e.g. Michael Dunn, *Australia and the Empire: From 1788 to the Present* (Sydney, 1984), 9-10; Manning Clark, *A Short History of Australia* (New York, 1969); R. W. Connell and T. H. Irving, *Class Structure in Australian History*, 109-10; Humphrey McQueen, *A New Britannia* (Melbourne, 1975).

10. W. D. Hussey, *The British Empire and Commonwealth 1500 to 1961* (Cambridge, 1968), chapter 21.

11. McNaughtan, 'Colonial Liberalism', 129-39; A. J. Reitsma, *Trade Protection in Australia* (Leiden, 1960), 45; Dunn, *Australia and Empire*, 64-5.

12. Charles Dilke, *Problems of Greater Britain* (London, 1890, p. 496), quoted in McNaughtan, 'Colonial Liberalism', 111.

13. R. A. Gollan, 'Nationalism, the Labour Movement and the Commonwealth,' in Greenwood (ed.), *Australia*, 145-95, 151.

14. L. G. Churchward, *Australia and America, 1788-1972* (Sydney, 1979), 97.

15. Argued by Hudson and Sharp, *Australian Independence*.

16. Geoffrey Sawer, 'Federalism', in Sol Encel, Donald Horne and Elaine Thompson (eds), *Change the Rules: Towards a Democratic Constitution* (Melbourne, 1977), 10-15.

17. K. W. Robinson, 'The Geographical Context of Political Individualism, 1860-1914', in Powell and Williams (eds), *Australian Space*, 231.

18. N. G. Butlin, A. Barnard, J. J. Pincus, *Government and Capitalism* (Sydney, 1982).

19. G. J. R. Linge, *Industrial Awakening: A Geography of Australian Manufacturing 1788 to 1890* (Canberra, 1979).

20. Reitsma, *Trade Protection*; G. D. Patterson, *The Tariff in the Australian Colonies, 1856-1900* (Melbourne, 1968).

21. See statement by David Syme, editor of Melbourne newspaper *The Age*, quoted in Reitsma, *Trade Protection*, 8.

22. Klaus Boehm, *The British Patent System: Administration* (Cambridge, 1967), 1-3. Also S. C. Gilfillan, *Invention and the Patent System* (Washington, 1964). For a history of the patent system see Fritz Machlup, *An Economic Review of the Patent System* (Washington, 1958), chapter 2.

23. Barton Hack, 'A History of the Patent Profession in Colonial Australia', Annual Conference of Institute of Patent Attorneys in Australia, Brisbane, 1984.
24. *NSWPD*, 15 June 1887, 2103-6.
25. Machlup, *Patent System*, 7.
26. Gilfillan, *Invention*, 14.
27. Dyster and Meredith, *Australia*, 32.
28. E. Dunsdorfs, *The Australian Wheat-Growing Industry* (Melbourne, 1956), 117-33.
29. C. Forster, 'Economies of Scale and Australian Manufacturing', in C. Forster (ed.), *Australian Economic Development in the Twentieth Century* (London and Sydney, 1970), 141. On manufacturing generally, see Linge, *Industrial Awakening*.
30. N. G. Butlin,'Company Ownership of NSW Pastoral Stations, 1865-1900', *Historical Studies of Australia and New Zealand*, 1950, 89-111.
31. J. W. McCarty, 'British Investment in Western Australian Gold Mining 1894-1914', *University Studies in History*, 1961-62, 7-23; Geoffrey Blainey, *The Rush That Never Ended: A History of Australian Mining* (Melbourne, 1963), 1978 edn, 100, 250.
32. R. T. Appleyard and C. B. Schedvin, 'Introduction', in Appleyard and Schedvin (eds), *Australian Financiers* (Melbourne, 1988), 1-10; Brian Fitzpatrick, *The British Empire in Australia* (Melbourne, 1969), 399-401.
33. W. A. Sinclair, *The Process of Economic Development in Australia* (Melbourne, 1976), 99; Margot and Alan Beever, 'Henry Gyles Turner', in Appleyard and Schedvin (eds), *Australian Financiers*, 108-36. For an opposing view, see Peter Cochrane, *Industrialization and Dependence* (Brisbane, 1980), 57.
34. A. G. L. Shaw, *The Economic Development of Australia*, 7th edn (Melbourne, 1980), 89.
35. Sinclair, *Economic Development*, 182-3; Cochrane, *Industrialization*, 35-6, 56-63; Appleyard and Schedvin (eds), *Australian Financiers*, 4.
36. S. B. Saul, *Studies in British Overseas Trade, 1870-1914* (Liverpool, 1960), 98-105. Saul is widely cited by Cochrane in support of the dependency argument.
37. Cochrane, *Industrialization*, 22-3; Stephen Hill, *The Tragedy of Technology* (London, 1988), 166-7.
38. N. G. Butlin, 'Perspectives of Economic Development' in Forster (ed.), *Australian Economic Development*, 266-327.
39. Linge, *Industrial Awakening*, 1.
40. Peter Milner, 'Gold Mining and the Development of Engineering Firms in Victoria', *Journal of the Royal Historical Society of Victoria*, 1986, 11-22, quote on p. 14.
41. Provided in M. S. Churchward and P. Milner, 'The Principal Engineering Establishments in Victoria in the Period 1842-1945', 2 vols, Department of Mechanical and Industrial Engineering, University of Melbourne, March 1988.

42. Twopeny, *Town Life*, 194.
43. Milner, 'Gold Mining', 20-1.
44. In particular, Don Stewart, 'The Heavy Engineering Industry and Engineering Products', in D. Fraser (ed.) *Sydney: From Settlement to City* (Sydney, 1989), 203-18.
45. Dyster and Meredith, *Australia*, 49.
46. Dunn, *Australia and Empire*, 72.
47. A fuller account and bibliography of the material presented here may be found in Jan Todd, 'Science at the Periphery: An Interpretation of Australian Scientific and Technological Dependency and Development Prior to 1914', *Annals of Science*, 1993, 33-58.
48. John Tebbutt, *The Sydney Observatory and the 'Sydney Morning Herald': A Plea for Astronomy in New South Wales* (Sydney, 1891).
49. S. G. Kohlstedt, 'Australian Museums of Natural History: Public Priorities and Scientific Initiatives in the 19th Century', *Historical Records of Australian Science*, 5 (4) (1983), 1-29.
50. On Austrians, see Marlene J. Norst and Johanna McBride, *Austrians and Australia* (Sydney, 1988).
51. Analysis based on lists of publications exchanged by the Royal Society of NSW, as published in its journal from 1876.
52. Derek Whitelock, *The Great Tradition: The History of Adult Education in Australia* (Brisbane, 1974), 98.
53. See Jan Todd, 'Colonial Adoption: the Case of Australia and the Sydney Mechanics' School of Arts', in Ian Inkster (ed.), *The Steam Intellect Societies* (Nottingham, 1985), 105-30. More generally, see Phillip Candy and John Laurent (eds), *Pioneering Culture: Mechanics' Institutes and Schools of Arts in Australia* (Adelaide, 1994).
54. For further details see Ian Inkster and Jan Todd, 'Support for the Scientific Enterprise, 1850-1900', in R. W. Home (ed.), *Australian Science in the Making* (Melbourne, 1988), 102-32.
55. Stephen Murray-Smith, 'Technical Education in Australia: A Historical Sketch', in E. L. Wheelwright (ed.), *Higher Education in Australia* (Melbourne, 1965), 170-91.
56. The following details come from Ann Moyal, 'Invention and Innovation: The Historian's Lens', *Prometheus*, 5 (1) (1987), 92-109, and Ian Inkster, 'Intellectual Dependency and the Sources of Invention', *History of Technology*, 12 (1990), 40-64.
57. Inkster, *ibid.*, 48.

3 MICROBES VERSUS POISONOUS PLANTS

1. A. Barnard, *The Australian Wool Market 1840-1900* (Melbourne, 1958), and 'A Century and a Half of Wool Marketing', in A. Barnard (ed.), *The Simple Fleece* (Melbourne, 1962), 475-89; N. G. Butlin, 'Distribution of the Sheep Population', in *ibid.*, 281-307.

2. N. G. Butlin, 'Growth of Rural Capital, 1860-1890', in Barnard, *Simple Fleece*, 322-39.

3. Gerard L. Geison, 'Pasteur', in Charles Gillespie (ed.), *Dictionary of Scientific Biography* (New York, 1974), vol. X, 350-416, 395.

4. For a description in English by a colleague of Pasteur, see Emile Roux, 'On Preventive Inoculation', *The Veterinarian*, 62 (1889), 632-5. For an early text see E. D. Klein, *Micro-organisms and Disease*, 3rd edn (London, 1896), 288-90.

5. For accounts of the process see, for instance, René Vallery-Radot, *The Life of Pasteur* (New York, 1924), esp. 311-13; René Dubos, *Pasteur: Free Lance of Science* (Boston, 1950), 330-1.

6. For emphasis on these points see Dubos, *Pasteur*, 342; Roux, 'Preventive Inoculation'.

7. For the relation of Pasteur's work to this debate see Geison, 'Pasteur', 380-90.

8. On Koch, see Claude E. Dolman, 'Koch', in Gillespie (ed.) *Dictionary*, vol. VII, 420-35. On Pasteur, Stephen Paget, *Pasteur and After Pasteur* (London, 1914), 60-61.

9. For an account of what Latour has called Pasteur's 'double strategy', see Bruno Latour, *The Pasteurisation of France* (Harvard, Cambridge, 1988), 67-72. For a detailed analysis (and bibliography) of Pasteur as a 'system builder' see J. H. Todd, 'Transfer and Dependence: Aspects of Change in Australian Science and Technology, 1880-1916', unpublished thesis, University of New South Wales, 1991, appendix I.

10. Geison, 'Pasteur', 385; G. L. Geison, 'Pasteur: A Sketch in Bold Strokes', in H. Koprowski and Stanley A. Plotkin (eds), *The World's Debt to Pasteur* (New York, 1985), 5-27. On the general support of French science see Robert Fox and George Weisz (eds), *The Organisation of Science and Technology in France 1808-1914* (Cambridge, 1980).

11. Emile Duclaux, *Pasteur: Histoire d'un esprit* (Paris, 1896), translated by E. F. Smith and F. Hedges as *Pasteur: The History of a Mind* (Philadelphia, 1920), 1973 edn, 225-6, 251-4.

12. Bruno Latour, 'Give Me a Laboratory and I Will Raise the World', in Karin D. Knorr-Cetina and Michael Mulkay (eds), *Science Observed: Perspective in the Social Study of Science* (Beverly Hills, 1983), 141-70.

13. Dr Lutaud, *M. Pasteur et la Rage* (Paris, 1887), cited in Geison, 'Pasteur: A Sketch', 14.

14. Dubos, *Free Lance*, 81; Hilaire Cuny, *Louis Pasteur: The Man and His Theories* (London, 1965), 20.

15. Duclaux, *Pasteur*, 293.

16. *Ibid.*, 187.

17. Vallery-Radot, *The Life*, 422-42.

18. W. A. N. Robertson, 'Milestones in the Pastoral Age of Australia', *Report of ANZAAS*, 1932, 295-325.

19. Commission to Enquire into Prevailing Epidemic Among Stock, 'First Report', in *NSW Government Gazette*, 11 March 1851, 440; 'Second Report', in *ibid.*, 17 June 1851, 938.

20. The basic structure for the future was contained in 'An Act for the Prevention and Cure of Diseases in Sheep', *Statutes of NSW*, 1863-64, 4-12.

21. W. T. Kendall, 'Twenty Years of the Veterinary Profession in Victoria', *Alma Mater*, June 1900, 123.

22. Edward J. McBarron, 'Life and Work of Alexander Bruce', unpublished manuscript, 1964, ML MSS 1231.

23. A. Bruce, 'Report on the Infectious and Contagious Diseases in Europe affecting or likely to affect the Live Stock in Australia', *VPLANSW*, 4 (1873-74), 955-65.

24. H. R. Seddon, *Diseases of Domestic Animals in Australia* (Canberra, 1955), vol. 1, part 5, 10.

25. Leslie Devlin, 'The Advent of Vaccination of Stock Against Anthrax in Australia', *AVJ*, 1943, 102-11; 'Report of the Australasian Stock Conference 1886', *JLCNSW*, 42 (2) (1887), 227.

26. James Cox, 'Presidential Address', *PLSNSW*, 1881-82, 866-9.

27. Anthony Willows, 'Report on Anthrax, or Cumberland Disease', *VPLANSW*, 34 (3) (1883), 957-62.

28. Devlin, 'The Advent', 104.

29. Edward Stanley, 'Report to Alexander Bruce', 17 December 1885, in Stock and Brands Branch, *Annual Report*, 1885, 15-16.

30. Edward Stanley, 'The Anthrax Disease' in NSW Anthrax Board, *Report on Experiments Demonstrating the Efficacy of Pasteur's Vaccine of Paris* (Sydney, 1889), 8-20, 11.

31. Alexander Bruce, Minute Paper to Under-Secretary for Mines, 24 December 1885, in Stock and Brands Branch, *Annual Report*, 1885, 16.

32. E. Stanley, 'Splenic Apoplexy in Sheep', in *Report of the Australasian Stock Conference* (Sydney, 1886), 17.

33. *Ibid.*

34. Stock Branch, *Annual Report*, 1886, 8.

35. W. M. Hamlet, Letter to Editor, *SMH*, 31 July 1918.

36. *NSWPD*, LA, 31 May 1887.

37. Alexander Bruce, Letter to Editor, *SMH*, 18 August 1888; *PLSNSW*, 1886, 908, 924.

38. Stanley, 'The Anthrax Disease'.

39. Devlin, 'The Advent', 104-5.

40. T. Coghlan, *Wealth and Progress of NSW*, 1892, 381.

41. F. Abigail, Letter to the Editor, *APR*, 15 March 1892, 544.

42. *The Times*, London, 25 February 1888, 15.

43. J. Creed, Letter to Editor, *Argus*, 19 March 1888.

44. Pasteur to *Le Temps*, 29 November 1887, and Pasteur to Sir Saul Samuel, 5 January 1888, printed in 'The Rabbit Pest', *JLCNSW*, 43 (4) (1887-88), 693.

45. Pasteur to Mrs Priestley, 30 December 1887, NLA MS 6530.

46. Jean Chaussivert, 'Letters regarding the "Pasteur Mission" in Australia', in Jean Chaussivert and Maurice Blackman (eds), *Louis Pasteur and the Pasteur Institute in Australia* (Sydney, 1988), 19.

47. Daniel Cooper to Henry Parkes, March 1888 (?), Parkes Correspondence, ML CYA 927, 139-40.
48. Devlin, 'The Advent', 105.
49. Pasteur to Adrien Loir, 26 February 1888, *PC*, BN; Stock Branch, *Annual Report*, 1887, 8.
50. Daniel Cooper to Parkes, 28 January 1888, Parkes Correspondence, ML CYA 920, 128-32.
51. Under-Secretary for Mines to Loir, 24 April 1888, ALCB I, AB, MS 100/1.
52. Devlin, 'The Advent', 105; A. Loir, *Pasteur's Vaccine of Anthrax in Australia* (Sydney, 1891), 3-4.
53. *SMH*, 4 May 1888, 4.
54. *SMH*, 12 May 1888, 4; *SMH*, undated, ALCB I; *SMH*, 11 May 1888, 4.
55. Cablegram, Agent-General to Parkes, 23 May 1888, CSC, AONSW, 44/3978.
56. *Queenslander*, 2 June 1888, 862.
57. *SMH*, undated, in ALCB I; Narrandera Stock Inspector to Loir, 14 June 1888, ALCB I; W. M. Hamlet, 'The History of the Anthrax Bacillus', in Anthrax Board, *Report*, 20-6, 23.
58. *Queenslander*, 16 June 1888, 921; unlabelled cutting, ALCB I, 85; *SMH*, 18 July 1888, 10.
59. *SMH*, undated, ALCB I, 87, 84.
60. *NSWPD*, 11 April 1889, 572-82; Devlin, 'The Advent', 105.

4 CONTAGION, CONFLICT AND COMPROMISE

1. Abigail, *APR*, 15 March 1892, 544; *Queenslander*, 6 November 1886, 737; Cooper to Parkes, 2 March 1888, Parkes Correspondence, vol. 7, 233-4 ML CYA 877; A. Loir, *A l'Ombre de Pasteur* (Paris, 1938), 109-10.
2. Adrien Loir, *La Microbiologie en Australie* (Paris, 1892), chapter 1; *NSWPD*, 12 October 1887, 463-70, 14 December 1887, 2109-24, 3 July 1888, 5995, 19 July 1888, 6711-1713; *APR*, 15 March 1892, 536, 15 September 1892, 817-20; *Queenslander*, 3 March 1888, 340.
3. Bruni, 'The Rabbit Plague', *Australasian*, 17 March 1888; letter from 'B', *SMH*, 4; *SMH*, 6 February 1888, 6.
4. Quoted by 'D'Englesqueville', *SMH*, 10 February 1888, 9.
5. NSW Board of Health, *Minutes*, 8, 15, 22, 29 February 1888; 'The Rabbit Pest', 12-5, in *JLCNSW*, 43 (4) (1887-88), 693.
6. A. Young, A. J. Sefton, N. Webb (eds), *Centenary Book of the University of Sydney Faculty of Medicine* (Sydney, 1984), 123.
7. Dr Wilkinson to Secretary for Lands, 9 February 1888, in 'The Rabbit Pest', 11.
8. Dr Weber to Editor, *Argus*, published in *Australasian*, 28 January 1888, 188; H. C. Wigg, 'On the Proposed Introduction of New Diseases into Australia', *JRSVic*, 1888, 28-33; *Australasian*, 24 March 1888, 636.
9. Agent-General to Parkes, 3 March 1888, CSC, AONSW 4/3978; *NSWPD*, 7 March 1888, (1887-88), 3133.

10. NSW Board of Health, *Minutes*, 9, 23 May 1888, AONSW 5/4936.
11. *NSW Government Gazette*, 1 (1888), 1993.
12. Loir, *l'Ombre*, 109-10; Pasteur to Loir, undated, probably late March 1888, Loir correspondence, private collection.
13. *Ibid.*
14. J. Creed to Editor, *Argus*, 19 March 1888.
15. Loir, *La Microbiologie*, chapter 1.
16. Rabbit Commission, *Progress Report and Minutes of Evidence*, IX-XII, 45 in *JLCNSW*, 47 (4) (1890), 775.
17. Pasteur to Loir, 1 March 1888, PC, BN.
18. Pasteur to Mrs Priestley, 17 June 1888, PC, BAS.
19. Quoted in *PR*, 15 August 1902, 395.
20. Pasteur to Jean-Baptiste, 14 June 1888, PC, BN; Rabbit Commission, *Progress Report*, XIII.
21. Pasteur to Mrs Priestley, 17, 23 June 1888, PC, BAS.
22. Pasteur to Salomons, 28 June 1888, PC, BN.
23. Pasteur to Loir, 11 July 1888, Loir Correspondence, private collection.
24. Germont and Loir to Editor, *SMH*, 1 April 1889; CSC, Special Bundle, 'Detention and opening of Pasteur Correspondence', AONSW 4/887.3; *NSWPD*, LC, 11 April 1889, 572-82.
25. Loir, *L'Ombre*, 112.
26. Dr R. C. Wigg to Editor, *SMH*, 9 August 1888; Bruce to Editor, *SMH*, 18 August 1888.
27. Referred to in Pasteur to Loir, 7 September 1888, PC, BN.
28. NSW Anthrax Board, *Report on Experiments Demonstrating the Efficacy of Pasteur's Vaccine of Paris* (Sydney, 1889).
29. Both published with Anthrax Board Report, along with resulting discussion.
30. *Proceedings of Australasian Stock Conference*, November 1889, 104, in *VPLAVic*, (1890) 2, 1413.
31. Pasteur to Germont and Loir, 2 August 1888, Loir Correspondence, private collection.
32. Loir, *L'Ombre*, 114.
33. *Ibid.*, 115.
34. Abigail to Parkes, 5 October 1888, Parkes Correspondence, ML A919, 118-20.
35. Colonial Secretary to Agent-General, 1 October 1888, and Agent-General to Colonial Secretary, 19 October 1888, CSC Special Bundle, AONSW 4/887.3.
36. Pasteur to Germont, via Agent-General, 2 December 1888, PC, BN; *Queenslander*, 22 December 1888, 1121.
37. Rabbit Commission, *Progress Report*, XXIII, appendix IV.
38. e.g. *NSWPD*, LC, 2, 28 November 1888.
39. *Transactions of Intercolonial Medical Congress of Australia*, Melbourne, 1889, 603.
40. Pasteur to Loir, 11 February, 11 March 1889, and Pasteur to NSW Agent-General, 9 March 1889, PC, BN.
41. Pasteur to Germont, 30 March 1889, PC, BN.

42. Pasteur to Loir and Germont, 7, 16 January 1889, PC, BN.
43. L. D. Devlin, 'The Advent of Vaccination of Stock Against Anthrax in Australia', *AVJ*, August 1943, 107.
44. Pasteur to Loir, 13 July 1889, Loir Correspondence, private collection; Pasteur to Grancher, 28 June 1889, PC, BN.
45. Loir and Germont to Bruce, and Germont to Devlin, 29 October 1889, in *Queenslander*, 21 December 1889.
46. *Proceedings of Australasian Stock Conference*, 1889.
47. *Echo*, 16 January 1890, 6.
48. *DT*, 9 January 1890; *Queenslander*, 18 January 1890.
49. Colonial Secretary to Agent-General, 21 January 1890, and Agent-General to Colonial Secretary, 4 February 1890, CSC, AONSW 4/3978-85.
50. Pasteur to Bruce, 24 April 1890, PC, ML DOC 2936.
51. *NSWPD*, 8 May 1890, 249; 5 June 1890, 1087.
52. *Echo*, 9 June 1890, 3.
53. *SMH*, 14, 24 June 1890; *Echo*, 28 June 1890, 7; unidentified cutting, ALCB III, private collection.
54. *NSWPD*, LA, 20 October 1892, 1308.
55. Loir, *L'Ombre*, 113-14.
56. Unidentified cutting, ALCB III, private collection.
57. A. Bruce, 'Treatment of Anthrax' in Report of Conference of Agricultural Societies, July 1890, in *Bulletin No. 3*, NSW Dept of Agriculture, Sydney, March 1891, 28-43.
58. Undated cutting, ALCB III, private collection.
59. e.g. *Echo*, *Evening News*, and *SMH*, 5 August 1890.
60. 'Pasteur's Vaccine', *JLCNSW*, 47 (2) (1890), 595; *Vaccinations Charbonneuses*, a pamphlet by the company, in ALCB I, AB, MS 100/1.
61. McGarvie Smith to Chief Inspector of Stock, 4 January 1912, ASF 109, AONSW 12/3522.
62. A. Devlin to Editor, *SMH*, 26 May 1890.
63. Unidentified cutting, ALCB III, private collection.
64. *Queenslander*, 1 December 1890.
65. A. Loir, *Pasteur's Vaccine of Anthrax in Australia* (Sydney, 1890).
66. Madame Pasteur to Loir, 17 January, 15 June 1891, Loir Correspondence, private collection.
67. e.g. *DT* and *Echo*, 18 September 1890.
68. *Echo* and *SMH*, 16 October 1890.
69. L. Devlin, 'The Advent', 109.
70. Stock and Brands Branch, *Annual Report*, 1890, p. 9 and Appendix G. Leslie Devlin, writing in 1943, quoted much higher figures but the Devlin records had long since been destroyed by white ants and there is no reason why figures obtained direct from Loir and Arthur Devlin would be understated.

5 FROM PARIS TO NARRANDERA

1. Stock and Brands Branch, *Annual Report*, 1890, p. 18, Appendix G; A. Loir, *Pasteur's Vaccine of Anthrax in Australia* (Sydney, 1890), 6.
2. Stock and Brands Branch, *Annual Report*, 1891, 22, Appendix I.
3. NSW Board of Health, *Minutes*, 8 March 1893, AONSW 5/4936-42.
4. L. D. Devlin, 'The Advent of Vaccination of Stock Against Anthrax in Australia', *AVJ*, August 1943, 102-11, 110.
5. Calculated from Gunn's 1891 estimate that treatment of 25,000 sheep cost £450, and vaccine for this number would have cost £346.
6. Gunn as 'Manager' to Editor, *Australasian*, 28 December 1891, and *SMH*, 19 December 1892.
7. 'Report of Conference of Agricultural Societies, July, 1890', in NSW Dept of Agriculture, *Bulletin No. 3*, March 1891, 38.
8. Bruce to Loir, 7 January 1892, ALCB II, AB MS 100/2; *APR*, 15 September 1892, 832, 846.
9. Quoted in *APR*, 15 March 1894, confirmed by annual statistics in *NSW Statistical Register*.
10. N. Cain, 'Pastoral Expansion and Crisis in New South Wales, 1880-1893: The Lending View', *Australian Economic Papers*, December 1963, 180-98.
11. Oscar Katz, 'Preventive Inoculation Against Animal Plagues', *Report of AAAS*, 1892, 692; Katz to Editor, *SMH*, 11,17,24 May 1890.
12. e.g. *NSWPD*, LC, 26 June 1890, 1638, 26 November 1891, 3029.
13. *NSWPD*, LA, 2 December 1891, 3259, 12 October 1892, 1069.
14. *NSWPD*, 1892, 2067, 1308, 1409, 1429; 'Ovis' to *Australasian*, 26 December 1891, 'Bovis' to *Australasian*, 17 September, 22 October, 19 November 1892.
15. Wilkinson to Editor, *SMH*, 11 November 1892, 11.
16. J. S. Horsfall to Goldsbrough Mort and Co., 13 July 1897, and J. S. Horsfall to Cooper, 9 June 1897, Goldsbrough Mort Papers, NBAC 2/704.
17. Richard Webster, *Bygoo and Beyond* (Ardlethan, 1955), chapter 11; J. W. Cunningham to *SSSJ*, 6 July 1923.
18. Alastair J. Gunn, 'The Honourable John Alexander Gunn', *Ballarat Historian*, 1985, 23-37.
19. Gunn, *ibid.*; Webster, *Bygoo*, chapter 11.
20. *ADB*, vol. 10 (Melbourne, 1988), 652-3; *Australian Encyclopedia*, vol. III (Sydney, 1965), 153.
21. Devlin, 'The Advent', 111; J. D. Stewart, 'The Advent of Vaccination Against Anthrax in Australia', *AVJ*, October 1943, 148-9.
22. Pasteur Anthrax Vaccine Laboratory of Australasia, *Anthrax and its Treatment* (Sydney, 1901).
23. *APR*, 16 July 1894, 221.
24. Anthrax Vaccine Committee, *Minutes*, 15 May, 2 July 1917, 2/7/1917, in ASF 109, AONSW 12/3522.
25. *Ibid.*, 2 July 1917.
26. R. Freweu to J. A. Gunn, 29 February 1896, ASF 109, AONSW 12/3522.

27. 'An Australian Expert', *Australasian*, 2 June 1894; Gunn to Editor, *APR*, 15 January 1899.
28. E. C. Pope to unidentified newspaper, in Gunn Cuttings, private collection; J. W. Cunningham to Editor, *SSSJ*, 6 July 1923.
29. e.g. Gunn as 'Manager' to Editor, *Australasian*, 10 October, 28 December 1891, 29 October, 19 December 1892.
30. Undated cutting, ALCB III, private collection.
31. W. C. Wilkinson to Editor, *SMH*, 19, 21 November, 7 December 1892.
32. Gunn as 'Manager' to Editor, *SMH*, 19 December 1892.
33. Gunn as 'Manager', 'Inoculation for Disease', *APR*, 15 May 1893.
34. McGarvie Smith to Chief Inspector of Stock, 4 January 1912, ASF 109.
35. *Australasian*, undated, in Gunn Cuttings, NBAC 55/1.
36. *APR*, 15 March 1894, 26, with excerpt from *Narrandera Argus*.
37. 'Bruni' article, *Australasian*, 2 June 1894.
38. Gunn to Goldsbrough Mort, 10 September 1893, Goldsbrough Mort Papers, NBAC 2/563/2.
39. W. D. Ferguson, *A Few Facts About Anthrax*, a pamphlet produced by Gunn's agent in Victoria, in Gunn Papers, NBAC 55/1.
40. *APR*, 16 July 1894.
41. D. McLeish to J. A. Gunn, 1 June 1894, ASF 109, refers specifically to provision of a testimonial in time for a coming meeting on the 4th.
42. Vaccination Returns of McGarvie Smith and Gunn, McGarvie Smith Institute Papers, ML MSS 2411/14-19.
43. Report on Retreat, 10 September 1896, p. 3, in Goldsbrough Mort Papers, NBAC 2/306/3.
44. Goldsbrough Mort, *Minutes*, 3 December 1895, NBAC 2B/36/1.
45. Gunn to *Town and Country Journal*, written April 1895, unlabelled cutting in Gunn Papers, NBAC 55/1.
46. *SSSJ*, 26 April 1895.
47. *Australasian*, 2 June 1894.
48. *APR*, 15 July 1895.
49. L. Devlin, 'The Advent', 110.
50. Torsten Hagerstrand, 'Quantitative Techniques for Analysis of the Spread of Information and Technology', in C. A. Anderson and M. T. Bowman (eds), *Education and Economic Development* (Chicago, 1965), 244-80.
51. *APR*, 15 July 1895.
52. H. R. Seddon, *Diseases of Domestic Animals in Australia*, vol. 1 part 5, (Canberra, 1955), 16.
53. D. Fletcher to Gunn, 2 June 1895, ASF 109.
54. W. Alison to Editor, *APR*, 15 August 1894.
55. Max Henry, 'The Incidence of Anthrax in Stock in Australia', *JRSNSW*, 1922, 44-61.
56. Loir, *Pasteur's Vaccine*, 10.
57. Seddon, *Diseases*, 34.
58. Calculated from *NSW Statistical Register* and the Vaccination Returns of McGarvie Smith and Gunn, ML MSS 2411/14-19.

59. Gunn as 'Manager' to Editor, *SMH*, 19 December 1892, 6.
60. McGarvie Smith and Gunn leaflets for various years are among the vaccination records of the firm.
61. Paul David drew attention to this kind of threshold in 'The Mechanisation of Reaping in the Ante-bellum Midwest', in H. Rosovsky (ed.), *Industrialization in Two Systems* (Wiley, 1966), 3-28.
62. e.g. J. S. Metcalfe, 'The diffusion of innovation: an interpretive survey', in G. Dosi, C. Freeman, R. Nelson, G. Silverberg, L. Soete (eds), *Technical Change and Economic Theory* (London, 1988), 560-89.

6 FROM FOREIGN TO DOMESTIC CAPABILITY

1. Bruno Latour, *The Pasteurisation of France* (Harvard, 1988), 100-10.
2. R. Dedonder, 'World-wide Impact of Pasteur's Discoveries and the Overseas Pasteur Institutes', in Hilary Kropowski and Stanley A. Plotkin (eds), *The World's Debt to Pasteur* (New York, 1985), 131-40.
3. *JRSNSW*, 1891, 46-51; *ibid.*, 89-92.
4. A. Loir and E. Stanley, 'The Susceptibility of the Kangaroo to Anthrax', *Agricultural Gazette of NSW*, 1891.
5. A. Loir and A. Bruce, 'Les Maladies du Betail en Australie', *Annales de l'Institut Pasteur*, March 1891, 177.
6. *Proceedings of the Australasian Stock Conference*, Melbourne, November 1889, 101 in *VPLAVic*, 1890, part 2, 1413.
7. 'Correspondence respecting establishment of an Australasian Stock Institute on Rodd Island', *VPLANSW*, 1892-93, 50, part 8, 587.
8. *Ibid.*
9. *Ibid.*, 9.
10. *APR*, 15 November 1892, 914.
11. *SMH*, 18 October 1892.
12. W. C. Wilkinson, *Transactions of Intercolonial Medical Congress of Australasia*, 1889, 609. Also Wilkinson to Editor, *SMH*, 19 November 1892, 11.
13. *NSWPD*, LA, 18 October 1892; 10 May 1893, 6911.
14. F. A. Blackman to Editor, *Brisbane Courier*, 21 October 1892.
15. Circular from Queensland Stockbreeders' and Graziers' Association, 14 December 1892, in ALCB II, AB MS 100/2.
16. *Brisbane Courier*, 3 December 1892.
17. *APR*, 15 March 1893; Loir to F. A. Blackman, May 1893, Loir Correspondence, private collection in Paris.
18. Blackman to Editor, *PR*, 15 November 1901.
19. *SSSJ*, 15 April 1898; Gunn to Editor, *APR*, 15 March 1898.
20. *SSSJ*, 6 July 1900, 4.
21. Personal papers of J. D. Stewart, Sydney University Archives, P35.
22. Stock and Brands Branch, *Annual Report*, 1900, 4, 21-2; *SSSJ*, 4, 9 July 1902.

23. Government Bureau of Microbiology, *Annual Report*, 1909.
24. *AVJ*, September 1951, 246-9; May 1960, 205-11.
25. J. D. Stewart, 'Presidential Address' to Veterinary Science Section of AAAS, 1913, copy in Stewart Papers, Sydney University Archives, P35.
26. H. R. Carne, 'Veterinary Research, 1900-1950, at the Veterinary School, University of Sydney', *AVJ*, September 1951, 258-61.
27. J. D. Stewart, Lectures Box, Stewart Papers, Sydney University Archives, P35.
28. Personal communication from Ian Gunn to writer, 30 June 1990.
29. Robert Webster, *Bygoo and Beyond* (Ardlethan, 1955), 86.
30. H. R. Seddon, *Diseases of Domestic Animals in Australia*, vol. 1 part 5, (Canberra, 1955), 33.
31. Gunn to Editor, *Town and Country Journal*, probably 1895, in Gunn Papers, NBAC 55/1.
32. L. Devlin, 'The Advent of Vaccination of Stock Against Anthrax in Australia', *AVJ*, August 1943, 110.
33. George Badgery to *Farmer and Settler*, 15 October 1918.
34. E. C. Pope to unidentified newspaper, 22 October 1918, Gunn Cuttings, private collection.
35. Letter, Gunn to Rougier, 19 January 1894, ASF 109, AONSW 12/3522.
36. Personal interview with Jack Gunn, a descendant.
37. NSW Board of Health, *Minutes*, 8 June, 29 July 1891, AONSW 5/4936-42.
38. NSW Department of Lands, *Annual Report*, 1891, 21.
39. NSW Board of Health, *Minutes*, 14 September 1892.
40. Letter reproduced in *APR*, 15 February 1892.
41. Gunn as 'Manager' to the *Australasian*, probably January 1894, in Gunn Papers, NBAC 55/1.
42. E. C. Pope letter, 22 October 1918, in Gunn Papers, NBAC 55/1.
43. Professor Anderson Stuart, 'Presidential Address', *JRSNSW*, 1894, 13.
44. Gunn to Goldsbrough Mort, 10 September, 2 November 1893, Goldsbrough Mort Papers, NBAC 2/563/2.
45. McMillan to Gunn, 14 February 1894, ASF 109.
46. Stuart, 'Presidential Address'.
47. *Australasian*, 2 June 1894.
48. N. Cain, 'Financial Reconstruction in Australia, 1893-1900', *Business Archives and History*, 1966, 166-83.
49. Transfer of mortgage document, 20 September 1879, and letter from Australasian Mortgage and Agency Co. to John Osborne, 1 October 1880, in Goldsbrough Mort Papers, NBAC 2/563/1 (1876-84). See also *ADB*, vol. 3, 506.
50. McGarvie Smith to Chief Inspector of Stock, 4 January 1912, ASF 109.
51. NSW Board of Health, *Minutes*, 20 May, 25 July, 14 August, 11 September, 2 October, 20 November 1894.
52. *SSSJ*, 26 April 1895.
53. *APR*, 15 July 1895, 254.
54. *SSSJ*, 26 April 1895.

55. Biographical sources include *ADB*, vol. 10, 652–3, and *Australian Encyclopedia* (Sydney, 1965), 153. See also Ursula Bygott, *A History of the McGarvie Smith Institute, 1918-1992*, Sydney University Monograph (Sydney, 1993), chapter 1, and interview with Smith in 'The Anthrax Scourge', *Sunday Times*, 27 July 1916.
56. *Sunday Times*, 27 July 1916, 5 May 1918.
57. Undated cutting, McGarvie Smith Institute Papers, ML MSS 2411/25X.
58. NSW Board of Health, *Minutes*, November–December 1890, and April–May 1891, AONSW, 5/4936-42; NSW Dept of Lands, *Annual Report*, 1891, 21.
59. H.J.D., 'The Biological Laboratory of Mr J. McGarvie Smith', *Sydney Mail*, 24 December 1892.
60. *JRSNSW*, 1891, 15.
61. *Transactions of Intercolonial Medical Congress of Australia*, 1892, 739.
62. 'Mr. McGarvie Smith's Report on the Ventilation of Sewers', *JLCNSW*, vol. 52, part 1 (1894), 734; *NSWPD*, 29 March 1894, 1878.
63. *SMH*, 10 March 1895.
64. *PLSNSW*, 1888, 400.
65. Devlin, 'The Advent', 110.
66. McGarvie Smith to Chief Inspector of Stock, 4 January 1912, ASF 109.
67. Webster, *Bygoo*, 86-7.
68. Francois Ray, *SSSJ*, 6 June 1923, and George Badgery, *Farmer and Settler*, 15 October 1918.
69. F. A. McKenzie, unidentified cutting, October 1918, in Gunn Cuttings, private collection.
70. J. R. Stewart, *SSSJ*, 15 October 1918.
71 McGarvie Smith to Chief Inspector of Stock, 4 January 1912, ASF 109.
72. Devlin, 'The Advent', 111.
73. Letter, McGarvie Smith to Chief Inspector of Stock, 4 January 1912, ASF 109.
74. Stuart to Jessie Gunn, 2 November 1918, ASF 109.
75. W. Hamlet to Francois Ray, 11 September 1918, Ray Papers, NLA MS 4662 Ray.
76. Webster, *Bygoo*, 86-7.
77. Gunn Trustees to unidentified newspaper, in cuttings in McGarvie Smith Institute Papers, ML MSS 2411/25X; draft letter, undated, from Marcus Gunn to a newspaper, and draft letter from Jessie Gunn to Editor of the *Australasian*, dated 13 May 1918, both in Gunn Papers, NBAC 55/1.
78. *APR*, 15 July 1895, 254.
79. McGarvie Smith to Editor, *SSSJ*, 18 June 1897.
80. *Anthrax and Inoculation*, pamphlet published in Melbourne, 1899, compiling newspaper correspondence between the firms of Pasteur and McGarvie Smith and Gunn, p. 5 (ML).
81. *Sunday Times*, 15 April 1918.
82. Anthrax Vaccine Committee, *Minutes*, 28 May 1917, ASF 109.
83. Anderson Stuart to Editor, *SMH*, 12 October 1918.

84. See J. H. Todd, 'Transfer and Dependence: Aspects of Change in Australian Science and Technology, 1880-1916', unpublished Ph.D. thesis, University of New South Wales, 1991, 207-19.
85. Statement by Dr Wall and Dr Tidswell, McGarvie Smith Institute Papers, McGarvie Smith Formula, 1919-55, ML MSS 2411/22, Item 3.
86. J. A. Gunn, 'Directions in the event of my death at Mr. Smith's hands for carrying on the vaccine', 28 March 1908, ASF 109.
87. G. L. Geison, 'Pasteur', in Charles C. Gillespie (ed.), *Dictionary of Scientific Biography*, vol. X (New York, 1974), 350-416.
88. L. Pasteur, 'Remarks on Anthracic Vaccination as a Prophylactic of Splenic Fever', *The Veterinary Journal*, 1882, 336-8.
89. D. M. and W. R. Klemm, 'A History of Anthrax', *Journal of the American Veterinary Medical Association*, vol. 135, July-December 1959, 458-62; C. F. Dawson, 'Anthrax with Special Reference to the Production of Immunity', US Bureau of Animal Industry, *Bulletin 137*, 1911.
90. McGarvie Smith to Editor, *SSSJ*, 18 June 1897.
91. *Australasian Medical Gazette*, 21 August 1899, 236-7.
92. Klemm and Klemm, 'A History', 461; G. S. Wilson and A. A. Miles, *Topley and Wilson's Principles of Bacteriology and Immunity*, 6th edn, 2092; Hutyra and Matek, *Special Pathology and Therapeutics of Diseases of Domestic Animals* (Chicago, 1912), 28-9.
93. Statement by Frank Wall and Frank Tidswell, undated, McGarvie Smith Institute Papers, ML MSS 2411/22, item 5.
94. McGarvie Smith to Chief Inspector of Stock, 9 July 1914, ASF 109.
95. e.g. A. Eichhorn, 'Experiments in Vaccination Against Anthrax', US Dept of Agriculture, *Bulletin No. 340*, 27 December 1915; H. C. Brown and E. W. Kirwan, 'Standardisation of Bacterial Suspensions by Opacity', *Indian Journal of Medical Research*, 1915, 763.
96. Australian Vaccine and Serum Co. See circular from agents, Hain and Searight, 13 November 1906, ASF 109.
97. McGarvie Smith to G. Pentland, 15 February 1899, Gunn Letterbook, NBAC 55/2.
98. G. Pentland to C. Lynne, May 1899, in unidentified news cutting, Gunn Papers, NBAC 55/1.
99. Gunn to R. Coldham, 23 August 1907, Gunn Letterbook, NBAC 55/2.
100. McGarvie Smith to Editor, *SSSJ*, 30 April 1901, 6.
101. Letter from Gunn, August 1905, Gunn Letterbook, NBAC 55/2.
102. *PR*, 15 October 1910, 870.
103. Gunn to H. Bradley, 11 May 1908, Gunn Letterbook, NBAC 55/2.
104. Gunn to lawyer, 12 November 1906, *ibid*.
105. Gunn, 'Directions in the event of my death'.
106. Minute paper by Chief Inspector of Stock, Mr Symons, 7 November 1913; George Valder to H. Healey, 10 November 1912, both in ASF 109.
107. H. Healey for Stockowners' Association to Under-Secretary of Agriculture, 1 December 1912, ASF 109.

108. J. Bate to J. M. Atkinson, 15 October 1913; H. Healey to Minister for Agriculture, 30 October 1913, ASF 109.
109. John Stewart to *Country Life*, 3 February 1925; Circular, John Stewart and Sons, 14 July 1916, ASF 109.
110. John Stewart to George Valder, 15 July 1916, ASF 109; *The Biological Institute of Australasia*, booklet by John Stewart and Sons, in ASF 109.
111. J. Stewart to McGarvie Smith, 14 July 1916; P. Parker to McGarvie Smith, 14 July 1916, both in ASF 109.
112. *Sunday Times*, 23 July 1916; G. Darnell-Smith, 'Memo on Mr McGarvie Smith's Anthrax Vaccine', 8 July 1916; Anthrax Vaccine Committee, *Minutes*, 8 May 1917, all in ASF 109.
113. Numerous papers relating to the establishment of the institute are in McGarvie Smith Institute Papers, ML. For a published account, see Bygott, *McGarvie Smith Institute*.
114. *Austral News*, June 1963, 1.
115. Bygott, *McGarvie Smith Institute*, 88.
116. Marcus Gunn to Dr Frank Tidswell, 19 December 1915, ASF 109.
117. Dept of Agriculture, 'Recognition of the Services of the late Mr J. A. Gunn', 27 October 1923, ASF 109.
118. Newspaper cuttings, Francois Ray Papers, NLA 4462 Ray.
119. Arthur Webster, *A World of Experience in Animal Health: Websters* (Sydney, 1988), 1-8; also personal interviews with Arthur Webster and his professional associates.

7 AUSTRALIAN GOLD, BRITISH CHEMISTS

1. E. S. Meade, *The Story of Gold* (London, 1908).
2. G. Blainey, *The Rush That Never Ended: A History of Australian Mining* (Melbourne, 1978), 78; J. Reynolds, *Men and Mines* (Melbourne, 1974), chapter 7.
3. C. G. W. Lock, *Practical Gold Mining* (London, 1889); T. Rose, *The Metallurgy of Gold* (London, 1894); Henry Louis, *A Handbook of Gold Milling* (London, 1894).
4. J. S. Macarthur, 'The Macarthur-Forrest Process of Gold Extraction', *JSCI*, 31 March 1890, 267-71.
5. Rose, *Metallurgy*; Macarthur, 'Macarthur-Forrest'.
6. *The Mineral Industry*, 33 (1924), 271.
7. Meade, *Gold*, 81; *The Economist*, 15 October 1881, 1283; T. P. Devine, 'The Gold Famine', *Victorian Review*, 5 (1882), 346-57. See also A. G. Kenwood and A. L. Lougheed, *The Growth of the International Economy, 1820-1960* (London, 1973), chapter 7.
8. *Report of the Board to Investigate Methods of Treating Pyrites*, Victorian Dept of Mines, 1874; *Report of the Royal Commission Appointed to Inquire into the Working of the Present Gold Fields Act and Regulations of New South Wales*, Sydney, 1871.

9. From figures in *NSW Statistical Register*.
10. Victorian Royal Commission on the Decline of Goldmining, *First Progress Report*, December 1890, 6.
11. *Ibid.*
12. Committee No. 14, 'The State and Progress of Chemical Science in Australia, with Special Reference to Gold and Silver Appliances', *Report of AAAS*, 1890, 283-92.
13. J. Macarthur, *JSCI*, 28 February 1898, 135.
14. James Park, *The Cyanide Process of Gold Extraction* (Auckland, 1897), 1.
15. Macarthur, 'Macarthur-Forrest', and 'Gold Extraction by Cyanide: A Retrospect', *JSCI*, 1905, 311.
16. Obituary, *JSCI Review*, 1920, 161.
17. S. G. Checkland, *The Mines of Tharsis* (London, 1967).
18. *Ibid.*, 123-6.
19. *London Mining Journal*, 20 June 1884.
20. Checkland, *Tharsis*, 127.
21. Alfred James, *Qld Government Mining Journal*, 23 (1922), 233; Report by Cassel Directors to Shareholders' Meeting, 29 October 1886, CCRO BM 20/20/1-3.
22. Cassel Board, *Minutes*, 9 November 1886, CCRO BM 20/14/1-6.
23. *Ibid.*, 12 November 1886.
24. Macarthur, 'Retrospect', 312.
25. J. E. Clennell, *The Cyanide Handbook* (New York, 1910), 25.
26. Cassel Board, *Minutes*, October-December 1887; *Local Industries of Glasgow*, British Association for the Advancement of Science, 1901, 170-2.
27. Macarthur, 'Retrospect', 312-14.
28. Clennell, *Cyanide Handbook*, 25-8.

8 TRANSFER AGENTS AND COLONIAL CONNECTIONS

1. *Report of Cassel Directors*, 1885, 2-10, CCRO BM 20/20/1-3. Also Cassel Board, *Minutes*, August-September 1885, CCRO BM 20/14/1-6.
2. *Report of Cassel Directors*, 1885, 7.
3. *ADB*, vol. 10, 263-4, 282-3.
4. *ADB*, vol. 5, 161-4; Thomas McIlwraith entry in D. B. Waterson (ed.), *Bibliographical Register of the Queensland Parliament 1860-1920* (Canberra, 1972).
5. *ADB*, vol. 5, 160-1.
6. K. Buckley and K. Klugman, *The History of Burns Philp* (Sydney, 1981), 22, Appendix A.
7. *Report of Cassel Directors*, 1885, 4-5.
8. *Ibid.*, 1888, 8-9.
9. Cassel Board, *Minutes*, 31 December 1885.
10. *Ibid.*, 9 April, 3, 10, 25 May 1888.

11. *Report of Cassel Directors*, 1888, 8-9.
12. Cassel Board, *Minutes*, 29 March, 9, 19 April 1888.
13. Warden's Report, Qld Dept of Mines, *Annual Report*, 1886, 34; 1888, 35.
14. Cassel Board, *Minutes*, 10 November, 31 December 1885.
15. *Ibid.*, November-December 1888, January 1889.
16. *Ibid.*, 31 January, 28 February 1889.
17. *Ibid.*, 28 February, 7 March 1889.
18. *Ibid*, 14 March 1889.
19. *Ibid*, 21 March 1889.
20. *Ibid*, 20 June 1889; *Report of Cassel Directors*, 1889, 4-5.
21. *Report of Cassel Directors*, 1889, 5-6.
22. Cassel Board, *Minutes*, 20 June 1889.
23. Chairman's Address, *Proceedings of Cassel AGM*, December 1889, 9, CCRO
    BM 20/20/1-3.
24. E. W. H. Fowles (ed.), *The Queensland Cyanide Case: The Australian Gold
    Recovery Company Ltd v The Day Dawn PC Gold Mining Company Ltd*
    (Brisbane, 1902), 31, 41-59.
25. Cassel Board, *Minutes*, 7 March 1889, also 1 November 1888 and 11 July
    1889.
26. H. C. Perry, *Memoirs of the Hon. Sir Robert Philp KCMG, 1851-1922* (Bris-
    bane, 1923), 125, 139-49, 170-2.
27. Buckley and Klugman, *Burns Philp*, 21, 30-1; Perry, *Philp*, chapter X; D. J.
    Murphy and R. B. Joyce (eds), *Queensland Political Portraits, 1859-1952*
    (Brisbane, 1978); C. A. Bernays, *Queensland Politics During Sixty Years,
    1859-1919* (Brisbane, 1919), 101.
28. Buckley and Klugman, *Burns Philp*, 21-2; Perry, *Philp*, 106-7.
29. W. Skinner (ed.), *The Mining Manual*, 1888, 256-7, 106-8; 1889-90, 227.
30. James Waters to Robert Philp, 8 September 1885, Philp Papers, OML,
    OM65.032, series 3/1013.
31. Andrew McIlwraith to James Burns, 15 October 1886, Philp Papers, OML,
    OM665.032, series 3/1016; Buckley and Klugman, *Burns Philp*, 29.
32. Cassel Board, *Minutes*, 11 July 1889.
33. *Ibid.*, 4 July 1889.
34. *Ibid.*, 11 July 1889.
35. *Ibid.*, 31 October 1889.
36. *Ibid.*, 31 October 1889, also 22 November 1889.
37. *Ibid.*, from 22 November 1889 to 8 February 1890.
38. George Fowler to Wolston Trubshawe, 22 August 1891, Letterbook 1, p.
    37, in George Swan Fowler Papers (GSF), Mortlock Library, Adelaide, BRG
    151/1.
39. Fowler to Trubshawe, 5 September 1891; Fowler to Philp, 3 December
    1891, Fowler to Malcolm McEacharn, 3 December 1891, all in GSF 1.
40. Andrew McIlwraith to Philp, 12 December 1890, Philp Papers, OML, OM
    65.032, series 3/1270.
41. *Report of Cassel Directors*, 1890, 5.
42. Fowles, *Cyanide Case*, 42.

43 Andrew McIlwraith to Philp, 12 December 1890, Philp Papers, 3/1270; Cassel Board, *Minutes*, 11 October 1890.
44. *Report of Cassel Directors*, 1891, 4.
45. *Ibid.*
46. Michael Cannon, *The Land Boomers* (Melbourne, 1977), 16, 50.
47. D. and J. Fowler, *Years to Remember: A Record of the First Hundred Years of the Business of D. and J. Fowler Limited* (Adelaide, 1954).
48. *Report of Cassel Directors*, 1891, 4.
49. Cassel Board, *Minutes*, May-August 1891.
50. Fowler to Duncan McIntyre, 29 October 1891, GSF 1.
51. Fowler to Trubshawe, 5 September 1891, GSF 1.
52. Fowler to W. Webster, 20 August 1891, GSF 1.
53. Fowler to Trubshawe, 15 December 1891, also 9 December 1891 and cables to London on 10 and 14 December 1891, all in GSF 1.
54. Fowler to Trubshawe, 15 December 1891, GSF 1.
55. *Ibid.*, 9, 15, 22 December 1891.
56. Cassel Board, *Minutes*, 22 December 1891, 5, 12 January 1892.
57. Fowler to Trubshawe, 22 August 1891, GSF 1.
58. *Ibid.*, 11 July 1893, GSF 2.
59. Buckley and Klugman, *Burns Philp*, 34-7.
60. *The Times*, 27 February 1892.
61. Cassel Board, *Minutes*, 19 January 1892.
62. *Mining Manual*, 1894, 27, 411.

9 A CHALLENGE FOR TECHNOLOGICAL IMPERIALISTS

1. George Fowler to Wolston Trubshawe, 5 September, 21 October 1891, Letterbook 1, George Swan Fowler Papers (GSF), Mortlock Library, Adelaide, BRG 151/1.
2. For details see J. H. Todd, 'Transfer and Dependence: Aspects of Change in Australian Science and Technology, 1880-1916', unpublished Ph.D. thesis, University of New South Wales, 1991, chapter 5.
3. W. R. Thomas, *TIMMet*, 1898-99, 155; Qld Royal Commission on Mining, *Report and Minutes of Evidence* (Brisbane, 1897), 60, 78; J. M. Maclaren, *TIMEng*, 1900-01, 395.
4. James Park, *The Cyanide Process of Gold Extraction* (London, 1904), 184.
5. Qld Royal Commission, *Report*, 270.
6. Fowler to Trubshawe, 21 June 1892, GSF 2.
7. *AMS*, 13 May 1893, 253.
8. Geoffrey Blainey, *The Rush That Never Ended: A History of Australian Mining* (Melbourne, 1978), 251-2.
9. Park, *The Cyanide Process*, 17. Chapter III is devoted to laboratory experiments.
10. Fowler to M. C. Solomon, 19 July 1894, GSF 3.
11. Fowler to Charles Fortiere, 18 October 1891, GSF 1.

12. *AMS*, 12 June 1889.
13. Vic. Royal Commission on the Decline of Goldmining, *Minutes of Evidence*, in *VPLAVic*, 5 (1891): Wagemann, p. 823, Dunn, p. 656.
14. *Ibid*, 117-21; Ballarat School of Mines, *Annual Report*, 1890, 50; 1891, 84.
15. Vic. Royal Commission, *Minutes*, 690; 'Rewards for the Discovery of New Goldfields', *Northern Miner*, 23 March 1891.
16. W. M. Hamlet, in *Report of AAAS*, 1892, 51, 557-61.
17. *Ibid.*, 1890, 283-92.
18. *Ibid.*, 1892, 57.
19. *Report of Cassel Directors*, 1891, 4, CCRO BM 20/20/1-3.
20. Blainey, *The Rush*, 86.
21. Warden's Report, Qld Dept of Mines, *Annual Report*, 1886, 34. These reports during the 1880s provide a picture of the field's activities.
22. Fowler to Trubshawe, 5 September, 15 December 1891, GSF 1; Cassel Board, *Minutes*, 10 December 1890, CCRO BM 20/14/1-6.
23. John Kerr, *Mount Morgan: Gold, Copper and Oil* (Brisbane, 1982).
24. Vic. Dept of Mines, *Annual Report*, 105-6.
25. Qld Dept of Mines, *Annual Report*, 1887, 48; Qld Royal Commission, *Report and Minutes*, 264-5; Fowler to Trubshawe, 22 August, 30 September 1891, GSF 1.
26. Fowler to Trubshawe, 5 September 1891; Fowler to Appleton and Wright, 2 October 1891, both in GSF 1.
27. Peter McIntyre to Fowler, 1 October 1891, GSF 1; Fowler to Mount Morgan, 2 November 1891, in *AMS*, 12 December 1891.
28. Fowler to Trubshawe, 7 October 1891, GSF 1.
29. *Ibid.*, 14 October 1891; Report by J. W. Hall, published in *AMS*, 17 October 1891, 6.
30. Diane Menghetti, 'Extraction Practices and Technology on the Charters Towers Goldfield', *North Australia Research Bulletin*, September 1982; William Lees, *The Goldfields of Queensland. Charters Towers Goldfield* (Brisbane, 1899); E. H. T. Plant, 'Gold Ores of Charters Towers: Their Nature and Treatment', *AMS*, 21 May 1900.
31. Duncan McIntyre to Fowler, 26 October 1891, GSF 1.
32. Fowler to Trubshawe, 5 September, 22 December 1891; Fowler to John McConnell, 31 October 1891, GSF 1; Fowler to Peter McIntyre, 29 January 1892, GSF 2.
33. Fowler correspondence, December 1891, GSF 1.
34. Fowler to Trubshawe, 15 December 1891; Fowler to Peter McIntyre, 17 November, 14 December 1891, GSF 1.
35. Fowler to Peter McIntyre, 2 March 1891; Fowler to Trubshawe, 2 March 1891, 17, 31 May 1892, GSF 2.
36. T. Coghlan, *Wealth and Progress of NSW*, 1890-91, 79.
37. 'Proceedings of Cassel AGM', December 1891, 9; Vic. Royal Commission, *Minutes*, 678-9, 818-19; Fowler to Trubshawe, 6 July, 22 August 1891, GSF 1, and 2 March 1892, GSF 2.

38. Vic. Dept of Mines, *Annual Report*, 1883-89; Ballarat School of Mines, *Annual Report*, 1890, 19.
39. Fowler to Trubshawe, 7 October, 22 August 1891, and Fowler to John McConnell, 30 October 1891, GSF 1.
40. Fowler to Trubshawe, 22 October 1891, GSF 1.
41. Duncan McIntyre to Fowler, 2 November 1891, GSF 1.
42. Fowler correspondence, November-December 1891, GSF 1.
43. Fowler to Trubshawe, 9 December 1891, GSF 1; Vic. Royal Commission, *Minutes*, 789-91.
44. Fowler to Burrall, 1 December 1891; Fowler to McEacharn, 3 December 1891, GSF 1.
45. Fowler to Trubshawe, 20 January, 16 February, 15 March 1892; Fowler to Burrall, 29 February 1892, GSF 2.
46. Fowler to Trubshawe, 6 April, 23 May 1892, GSF 2.
47. D. and J. Fowler, *Years to Remember: A Record of the First Hundred Years of the Business of D. and J. Fowler Limited* (Adelaide, 1954); John Price, *Memoir of George Swan Fowler* (Adelaide, 1897); *ADB*, vol. 4, 207.
48. *SAPD*, 6 October 1886, 1206-8, 15 November 1888, 1750; G. D. Combe, *Responsible Government in South Australia* (Adelaide, 1957), 115-20; H. J. Burgess (ed.), *The Cyclopedia of South Australia* (Adelaide, 1909), 128-9.
49. Fowler to Terry, 12 August 1891; Fowler to Peter McIntyre, 13 August 1891; Fowler to Trubshawe, 19 August 1891, GSF 1.
50. Fowler to Trubshawe, 7 October 1891, GSF 1; 2 February 1892, GSF 2.
51. Fowler to Burrall, 15, 21 March 1892; Fowler to Trubshawe, 6, 12 April, 17 May 1892, GSF 2.
52. *Sands and McDougall's Directory of South Australia* (Adelaide, 1890), 936; *AMS*, 8 August 1891, 14; Fowler to Duncan McIntyre, 31 December 1891, GSF 1.
53. Cassel Board, *Minutes*, 10 July 1890; Fowler to Trubshawe, 19 August 1891, GSF 1.
54. Duncan McIntyre to Fowler, 18 January 1892; Fowler to Rosewarne, 28 January 1892, GSF 1; Rosewarne to Editor, *SA Register*, published in *AMS*, 17 February 1894, 99.
55. Fowler to Trubshawe, 15, 29 December 1891, GSF 1.
56. Fowler to R. Caldwell, 2 November 1891, GSF 1.
57. Fowler to Peter McIntyre, 20 October 1891; Fowler to Burrall, 20 October 1891, GSF 1
58. J. V. Parkes, *Report on Northern Territory Mines and Mineral Resources* (Adelaide, 1892), 30.
59. *SAPD*, 11 October 1893, 2076.
60. *AMS*, 16 December 1893, 716; 27 January 1894, 54.
61. Cassel Board, *Minutes*, 13 August 1895.
62. Fowler to Webster, 20 August, 2 November 1891; Fowler to Trubshawe, 22 August 1891, GSF 1.
63. Fowler to R. Caldwell, 2 November 1891; Fowler to A. Higham, 22 December 1891, GSF 1.

64. Fowler to Duncan McIntyre, 20 October 1891; Fowler to Trubshawe, 7 October 1891, GSF 1.
65. Fowler to Trubshawe, 20 August 1892, GSF 2.
66. Fowler to J. McConnell, 18 November 1891; Fowler to Peter McIntyre, 24 November 1891, GSF 1.
67. Fowler to Trubshawe, 7 September 1892, GSF 2.
68. *Ibid.*, 22 March 1892, GSF 2.
69. E. W. H. Fowles (ed.), *The Queensland Cyanide Case* (Brisbane, 1902), 29.
70. Fowler to son, 7 June, 5 July 1894, GSF 3.
71. *Ibid.*, 13 July 1893.
72. Fowler to A. A. Blackman, 2 October 1891; Fowler to Duncan McIntyre, 17 November 1891; Fowler to Trubshawe, 9 December 1891, GSF 1.
73. Fowler to Trubshawe, 2 February, 21 June 1892; Fowler to W. Webster, 30 January 1892, GSF 2.
74. Fowler to Trubshawe, 31 May 1892; *SA Register*, 9 July 1895; Cassel Board, *Minutes*, 5 November 1895.
75. Gordon Wilson to Under-Secretary for Mines, 7 August 1896, in 'Saving of Gold by the Cyanide Process', *QldPVP*, 4 (1896), 273.
76. *Engineering and Mining Journal*, 18 January 1890, 82-3.
77. *Ibid.*, 6 August 1892, 121.
78. *AMS*, 6 August 1892, 72, 19 November 1892.
79. Fowler to Trubshawe, 7 September 1892, GSF 2; *AMS*, 19 November 1892; Cassel Board, *Minutes*, 3 December 1892, 28 March 1893.
80. *AMS*, 9 December 1893, 15 January 1894, 15.

10 GOVERNMENTS, EXPERTS AND INSTITUTIONAL ADJUSTMENT

1. Text of Romer judgment in *AMS*, 22 December 1894, 735.
2. T. I. Williams and S. Withers (eds), *A Biographical Dictionary of Scientists* (London, 1974); *Who's Who in Mining and Metallurgy* (London, 1910).
3. Quoted in text of judgment of Transvaal case, *AMS*, 21 January 1897, 1610-13, see 1613.
4. Text of judgment of British Appeal Court, *AMS*, 25 May 1895, 297; 1 June 1895, 310-11.
5. W. A. Dixon, 'On a Method of Extracting Gold, Silver, and Other Metals from Pyrites', *JRSNSW*, 1877, 93-111.
6. Quoted by Justice Smith in his judgment, *AMS*, 25 May 1895, 297.
7. *AMS*, 17 November 1894, 600, 666. Also 8 September 1894, 526.
8. *AMS*, 24 November 1894, 678.
9. George Fowler to W. S. Paull, 13 October 1895; Fowler to son, 19 October, 28 December 1894, Letterbook 3, George Swan Fowler Papers (GSF), Mortlock Library, Adelaide, BRG 151/1.
10. *AMS*, 22 December 1894, 733.
11. *AMS*, 23 March 1895, 168.
12. Fowler to M. C. Solomon, 21 June, 19 July, 3 September, 23 October 1894 GSF 3.

13. *AMS*, 30 January 1896, 788, 21 January 1897, 1612. Also Fowler to son, 12 July 1894; Fowler to William Murray, 8 January 1895, GSF 3.
14. *AMS*, 20 April 1895, 224.
15. *QldPD*, 27 September 1895, 1075-88.
16. *AMS*, 18 August 1894, 482; 15 December 1894, 719; 18 May 1895, 289; 15 April 1896, 1002; 30 September 1895, 2272-3.
17. Fowler to John Macarthur, 20, 21 December 1894, 23 February 1895; Fowler to Charles McCulloch, 22 December 1894, GSF 3. Also *AMS*, 18 May 1895, 289; 21 January 1897, 1610.
18. Gordon Wilson to James Fowler, 21 February 1895, in Personal Papers of James Richard Fowler, 1890-1922, Mortlock Library, Adelaide, PRG 34/6.
19. Fowler to A. McCulloch, 23 April, 12 June 1895, GSF 3.
20. *QldPD*, 27 September 1895, 1073.
21. *AMS*, 10 August 1895, 452.
22. *QldPD*, 27 September 1895, 1088.
23. *Ibid.*, 1078.
24. Petition in *AMS*, 10 August 1895, 452.
25. e.g. *QldPD*, LA, 23 August, 27 September 1895, 1073.
26. Quote in *QldPD*, 27 September 1895, 1084. For full range of government argument, see pp. 1076-88.
27. *QldPD*, 23 June 1896, 73; 19 August, 29 October 1896, 1396; 'Saving of Gold by the Cyanide Process', *QldPVP* 4 (1896), 273.
28. *AMS*, 19 May 1898, 2931.
29. Fowler to son, 10 August 1894, GSF 3; SA Crown Lands Dept to Qld Under-Secretary of Mines, 13 July 1896, in *QldPVP*, 4 (1896), 273.
30. *SAPD*, 20 December 1893, 3739, 13 June, 15 November 1894, 2352; *AMS*, 22 December 1894, 733.
31. *AMS*, 11 May 1895, 265; *SAPD*, 12 June 1895, 565, 4 July 1895, 415, 23 July 1895, 652.
32. *SAPD*, 7 September 1898, 436-7.
33. *SAPD*, 8 July 1896; 29 October 1896, 639; 22 July 1897, 247.
34. *SAPD*, 24 June 1897, 59; 7 September 1898, 434-8.
35. *SAPD*, 29 October 1896, 639; 24 June 1897, 59; 22 July 1897, 247.
36. Memo by Sir John Forrest, in *AMS*, 23 June 1898 and *The Times*, 14 November 1900, 14b.
37. *AMS*, 23 June 1898; *Western Argus*, 3 November 1898, 10.
38. WA Royal Commission on Goldmining, *Report and Evidence*, in *WAMVP*, 1, 2 (1898), Paper 26.
39. *AMS*, 5 May 1898, 2888, 12 May 1898, 2912. Also *Western Argus*, 20 October 1898, 15; 3 November 1898, 10.
40. *The Times*, 10 December 1900, 130.
41. *AMS*, 10 January 1901, 130, 25 April 1901.
42. *WAPD*, 16 October 1900, 1025.
43. *Ibid.*
44. *Ibid.*, 1025.
45. *Ibid.*, 27 November·1900, 1908-9, 1029.

46. *Ibid.*, 1929.
47. Petition from Kalgoorlie and Coolgardie Chambers of Mines, and Petition from AGRC, *WAMVP*, Session 6, 1900, vol. II, 15 August to 5 December, A16, A18.
48. *AMS*, 15 June 1896, 1083-4; 9 July 1896, 1162.
49. Barton Hack, *A History of the Patent Profession in Colonial Australia*, Annual Conference of Patent Attorneys of Australia, Brisbane, 1984, 13-4.
50. Text of judgment, *AMS*, 15 June 1896, 1083-4.
51. *AMS*, 9 July 1896, 1162-3.
52. e.g. James MacTear, 'On the So-called "Selective Action" of Very Dilute Solutions of Cyanide of Potassium', *TIMMet*, 1895-96, 37-49.
53. Text of judgment, *AMS*, 21 January 1897, 1610-13, 28 January 1897, 1628-9.
54. Quoted in G. G. Turri, 'Recent Developments in the Attempt to Amend the Cyanide Patent', *TAusIMEng*, 1897, 195-213.
55. *AMS*, 8 July 1897, 2014.
56. Text of Holroyd judgment, *AMS*, 22 December 1898.
57. 'Macarthur and Forrest's Patent', *Argus Law Reports*, vol. V, 16 May 1899, 89-100.
58. *Ibid.*, 94.
59. 'The Cyanide Litigation', *AMS*, 20 April 1899.
60. *AMS*, 31 August 1899, 163; 14 September 1899, 210; 28 September 1899, 242; 5 October 1899, 261, 270; 19 October 1899, 305.
61. *VicPD*, 17 October 1899, 1921-34, 2087; *AMS*, 30 November 1899, 415; 3 February 1900, 116.
62. *AMS*, 26 October 1899, 320; 30 November 1899, 415; 25 January 1900, 71. Also *Daily Telegraph*, 14 November 1899, 4.
63. *VicPD*, 1899, 3024.
64. *VicPD*, 14 February 1900, 3866-910. Also *AMS*, 1 February 1900, 94; 8 February 1900, 116; 22 February 1900, 184; 3 May 1900, 396.
65. *AMS*, 22 November 1905, 490.
66. *AMS*, 15 October 1896, 1382-3.
67. *NSWPD*, 86 (1896). For Want see pp. 4563-4, but also 3905, 4562-3.
68. *Ibid.*, 4565.
69. *AMS*, 15 October 1896, 1382-3.
70. Documents in 'Amendments to Cyanide Patent No. 453', in Attorney-General's Special Bundle, AONSW, 5/4706.
71. *AMS*, 25 March 1897, 16 November 1899.
72. Want and Smith objections in Attorney-General's Special Bundle, AONSW, 5/4706.
73. *The Bulletin*, 17 July 1897, 7; 24 July 1897, 7.
74. Henry Deane, 'Anniversary Address', *JRSNSW*, 1898, 27-8. Also *NSWPD*, 22 July 1897, 2230; 4 August 1897, 2580; 5 October 1897, 3548; 7 October 1897, 3672; 8 December 1898, 3074-84; *AMS*, 12 August 1897, 2107.
75. Declarations are in Attorney-General's Special Bundle, AONSW 5/4706.
76. *AMS*, 22 December 1898, 3545; Turri, 'Recent Developments', 211-12.

77. William Skey, *Transactions New Zealand Institute*, vol. VII, 379; *AMS*, 14 October 1897, 2318.
78. *The Bulletin*, 24 June 1899, 11; *AMS*, 29 June 1899, 22 November 1905; *NSWPD*, 15 November 1899, 2233, 21 November 1899, 2363-4; *The Mineral Industry*, vol. XI, New York, 1903, 334.
79. *Qld Government Mining Journal*, 15 April 1901, 160.
80. Cassel Board, *Minutes*, 11 July 1889, CCRO BM 20/14/1-6.
81. *Qld Government Mining Journal*, 14 June 1902, 304.
82. J. S. Macarthur to J. R. Fowler, 22 February 1895, James Fowler Papers, Mortlock Library, Adelaide, PRG 34/6.
83. E. W. H. Fowles (ed.), *The Queensland Cyanide Case* (Brisbane, 1902).
84. W. A. Dixon, 'Note on the So-called "Selective Action" of Cyanide of Potassium for Gold', *TIMMet*, 1897-98, 88-93, and *The Australian Technical Journal*, 28 March 1898, 54-60.

11 FROM GLASGOW TO KALGOORLIE

1. Cassel Board, *Minutes*, 23 February, 20 October 1897, CCRO BM 20/14/1-6.
2. *Qld Government Mining Journal*, 14 November 1903, 589.
3. *AMS*, 18 August 1894, 431.
4. AGRC Chairman's Address, *The Times*, 19 March 1897.
5. *AMS*, 7 December 1897, 2477.
6. Qld Royal Commission on Mining, *Report with Minutes of Evidence* (Brisbane, 1897), 58, 103.
7. G. Beilby, *JSCI*, 28 February 1898, 132.
8. Tom Coventry, *The Existing Conditions of Queensland Mining* (Brisbane, 1899), 7.
9. Qld Dept of Mines, *Annual Report*, various years.
10. Calculated from figures in R. P. Rothwell (ed.), *The Mineral Industry*, vol. VII, 1898, 297-8.
11. *AMS*, 31 May 1900, 494-5.
12. *AMS*, 16 December 1893, 715-16; 9 June 1894, 339.
13. *SAPD*, LA, 11 October 1893, 2076.
14. SA Dept of Mines, *A Short Review of Mining Operations in the State of South Australia*, 1903; 1904.
15. *AMS*, 14 April 1894, 224; 1 September 1894, 511.
16. *AMS*, 6 February 1896, 814; 7 January 1897, 1575; Karl Schmeissner, *The Goldfields of Australasia* (London, 1898), 196; NSW Dept of Mines, *Annual Report*, several years.
17. W. R. Thomas, *TIMMet*, 1898-99, 145-56; *AMS*, 5 August 1897, 2086.
18. Estimated from figures in NSW Dept of Mines, *Annual Report*, and *NSW Statistical Register*, various years.
19. Vic. Dept of Mines, *Annual Report*, 1893, 7, 16-17.
20. *AMS*, 15 July 1897, 2028.

21. Vic. Dept of Mines, *Annual Report*, various years; *AMS*, 19 July 1900.
22. Vic. Dept of Mines, *Annual Report*, 1895, 26-7.
23. *Victorian Year Book*, 1904, 545.
24. *AMS*, 14 January 1897, 1586.
25. *AMS*, 22 February 1900, 184.
26. Vic. Dept of Mines, *Annual Report*, 1905, 20.
27. Fowler to son, 24 June, 13 July 1894, GSF 3.
28. For details of estimation see J. H. Todd, 'Transfer and Dependence: Aspects of Change in Australian Science and Technology, 1880-1916', unpublished Ph.D. thesis, University of New South Wales, 1991, appendix 5.
29. K. S. Blaskett, 'Ore Treatment at Western Australian Gold Mines', *Chemical Engineering and Mining Review*, 10 November 1951, 43.
30. W. A. Prichard and H. C. Hoover, in *The Mineral Industry*, vol. XI, 1902, 338.
31. WA Dept of Mines, *Annual Report*, 1897, 19-21.
32. J. M. McClaren, *TAusIMEng*, 1900-01, 399-400.
33. E. H. T. Plant, 'Gold Ores of Charters Towers', *AMS*, 31 May 1900, 494-5.
34. E. W. H. Fowles (ed.), *The Queensland Cyanide Case* (Brisbane, 1902); Qld Royal Commission, *Report and Evidence*, 100-15.
35. Gordon Wilson to Under-Secretary for Mines, 7 August 1896, in *QldVP*, 4(1896), 273-4.
36. Diane Menghetti, 'Extraction Practices and Technology on the Charters Towers Goldfield', *North Australian Research Bulletin*, 8, September 1982, 1-16; A. G. Charleton, *TFedIMEng*, 1892-93, 403-4: W. Blanc, in Qld Dept of Mines, *Annual Report*, 1900, 38.
37. *AMS*, 20 September 1899, 320.
38. Goyder's work is detailed in reports in SA School of Mines, *Annual Report*, 1893-96; in *TAusIMEng* 1893, 98; 1895, 159-170; and in *Transactions of the Royal Society of SA*, 1895, 25-6.
39. Cited by Goyder as coming from J. Macarthur, *Journal of Chemical Trade*, 1890, 269.
40. SA School of Mines, *Annual Report*, 1894, 145; *AMS*, 16 December 1893, 716, 27 January 1894, 54; *Report of Cassel Directors*, 1893, 6-7.
41. S. B. Christy, *TAmIMEng*, December 1896, 735-45.
42. G. B. Bodlander, *Zeits. fur angen. Chem*, 1896, 583-7.
43. *AMS*, 16 June 1894, 354.
44. J. S. Wells, *Engineering and Mining Journal*, 1895, 584-5.
45. Fowler to son (James), August 1894, 10 August 1894, GSF 3.
46. SA School of Mines, *Annual Report*, 1895, 166-71.
47. *SAPD*, 4 September 1895, 1256.
48. Calculated from figures in SA Dept of Mines, *Annual Report*, 1903; 1904.
49. *TAusIMEng*, 1895, 167.
50. W. B. Gray, *ibid.*, 1897, 138-45.
51. W. H. Gaze, *A Handbook of Practical Cyanide Operations* (Sydney, 1898), 9; J. E. Clennell, *The Cyanide Handbook* (New York, 1910), 293-4.
52. *AMS*, 13 April 1895, 213.

53. *AMS*, 30 March 1895, 182.
54. *AMS*, 10 February 1894, 88; 17 February 1894, 94; 12 May 1894, 283; 1 October 1896, 1356. Also *Report of AAAS*, 1892, 61; J. Storer, *Proceedings of the Engineering Association of NSW*, 1888; 1889.
55. *AMS*, 20 April 1895, 226-7.
56. *AMS*, 30 March 1895, 182; 27 April 1895, 238-9; 20 April 1894, 226-7; 27 April 1895, 238-9; 4 May 1895, 245; 25 May 1895, 294-5; 7 August 1895, 463-7.
57. *AMS*, 21 January 1897, 1601-2.
58. A. R. Canning, *TIMMet*, 1898-99, 268; *AMS*, 28 September 1899, 250.
59. *Australian Technical Journal of Science, Art and Technology*, 1897, 83-4, 31 December 1902, 63; *AMS*, 28 November 1901, 838.
60. C. A. Mulholland, *Engineering and Mining Journal*, 1 June 1895.
61. e.g. *TAmIMEng*, December 1896, 735-47; February-July 1897, 821-5.
62. Wilfred Macdonald, *TIMMet*, 1905-06, 526-39.
63. W. R. Thomas, *ibid.*, 1898-99, 145-56.
64. *AMS*, 15 October 1903, 514, 22 October 1903, 549; NSW Dept of Mines, *Annual Report*, 1905, 18; 1910, 19.
65. For estimation see Todd, 'Transfer and Dependence', chapter 6.
66. Vic. Dept of Mines, *Annual Report*, 1896; 'Lab Report', *ibid.*, 1897, 27.
67. *AMS*, 8 July 1897, 2014, 1 March 1900, 187; *ADB*, vol. 4, 115; *VicPD*, 14 February 1900, 3868.
68. *AMS*, 1 July 1897, 1994.
69. For Deeble's work also see Vic. Dept of Mines, *Annual Report*, 1896, 'Lab Report', *ibid.*, 1897, 27.
70. G. Turri, *TAusIMEng*, 1897, 209-11.
71. James MacTear, 'Notes on the South German Mine, Maldon, Victoria', *TIMMet*, 1897-98, 49.
72. W. B. Gray, *TAusIMEng*, 1897, 138-41.
73. MacTear, 'Notes', 51. See also Vic. Dept of Mines, *Annual Report*, 1887, 20-1; 1889, 33-4.
74. D. White and T. M. Simpson, 'Complete specification for an improved method and apparatus for extracting precious metals from slimes or other finely divided material containing the same', private copy.
75. 'Secretary's Report', in Vic. Dept of Mines, *Annual Report*, 1898, 1900; Fowles, *Cyanide Case*, 24.
76. Blaskett, 'Ore Treatment', 44; Chamber of Mines of WA, *Monthly Report*, 1903.
77. W. A. Prichard and H. C. Hoover, 'The Treatment of Sulpho-Tellurides at Kalgoorlie', *The Mining Industry*, 1902, 339; H. F. Bulman, 'The Kalgoorlie Gold-Mines, Western Australia', *TFedIMEng*, 1898-99, 537; George Bancroft, 'Kalgoorlie, Western Australia and its Surroundings', *TAmIMEng*, 1898, 97.
78. John Yates, 'The Macarthur-Yates Process of Gold Extraction', *TFedIMEng*, 1896-97, 36-2; Fowler to son, 2, 11 November 1894, 25 January 1895, GSF 3; Cassel Board, *Minutes*, 1 July 1895.

79. H. P. Woodward, *TIMMet*, 1897-98, 14-27; *Western Argus*, 22 March 1900, 17.
80. WA Royal Commission on Mining, *Report and Minutes of Evidence* (Perth, 1898), 391; *Western Argus*, 18 August 1898, 4.
81. Prichard and Hoover, 'Sulpho-Tellurides', 339.
82. D. Clark, *West Australian Mining and Metallurgy* (Sydney, 1904), 82; Alfred James, *Cyanide Practice* (London, 1901), 73; Ted Mayman, *The Mile that Midas Touched* (Sydney, 1964), 126; G. Blainey, 'Herbert Hoover's Forgotten Years', *Business Archives and History*, 1962, 58.
83. From figures in Bulman, 'Kalgoorlie', 360.
84. J. K. Wilson, 'The Filter-Press Treatment of Slime in Western Australia', in James, *Cyanide Practice*, 83-4.
85. Clark, *West Australian*, 44.
86. James, *Cyanide Practice*, 89.
87. *AMS*, 4 January 1900.
88. WA Royal Commission, *Report and Evidence*, 391, 436; W. A. Frecheville, *TIMMet*, 1897-98, 147.
89. A. James, 'Notes on the Treatment of Kalgoorlie Sulpho-Telluride Ores', *TIMMet*, 1899-1900, 484-99, and *Cyanide Practice*, 90-100; Clark, *West Australian*, 21, 54; *Western Argus*, 1 March 1900, 30.
90. W. E. Simpson, *TIMMet*, 1904-05, 22-59; F. G. Brindsden, 'Roasting Sulpho-Telluride Gold Ore', *TAusIMEng*, March 1910; Blaskett, 'Ore Treatment', 43-53.
91. H. Knutsen, *TIMMet*, 1902, 2-36.
92. Blaskett, 'Ore Treatment', 45-7.
93. *Western Argus*, 8 March 1900, 8, 22 March 1900, 17, 5 April 1900, 5; Clark, *West Australian*, 65; E. W. Nardin, 'Bromo-cyaniding of Gold Ores', *TAusIMEng*, 1907, 73-82.
94. James, *Cyanide Practice*, 94.
95. Clark, *West Australian*, 95, 165.
96. G. W. Williams, 'Metallurgy of the Kalgoorlie Goldfield', *Engineering and Mining Journal*, 15 February 1908, 349.
97. G. A. Richard, 'Statistics and Economics', *TAusIMEng*, 1901, 18, 29.

12 OUT OF THE HANDS OF 'RULE OF THUMB MEN'

1. *NSWPD*, 12 December 1888, 1109.
2. *NSWPD*, 27 March 1888, 3609-11.
3. *NSWPD*, 12 December 1888, 116-17; 4 June 1889, 1887-98.
4. Vic. Dept of Mines, *Annual Report*, 1897, 29; Vic. Royal Commission on the Decline of Goldmining, *Report and Evidence*, in *VPLAVic*, 5 (1891), 19, 43-5, 72-3, 78, 647, 710, 823; *SA Register*, 10 June 1889, 7; 'Report of the Board of Technical Education' (Adelaide, 1888), in *SAPPP*, 3 (1888), Paper 33.

5. C. A. Bernays, *Queensland Politics During Sixty Years, 1859-1919* (Brisbane, 1919), 365.
6. George Nadel, *Australia's Colonial Culture* (Melbourne, 1957); Derek Whitlock, *The Great Tradition* (Brisbane, 1974); Stephen Murray-Smith, 'A History of Technical Education in Australia . . . ', unpublished Ph.D. thesis, University of Melbourne, 1966.
7. W. Perry, *A History of the Ballarat School of Mines and Industries* (Ballarat, 1984); F. Cusack, *Canvas to Campus* (Melbourne, 1973); Vic. Royal Commission, *Report and Evidence*, 4, 256.
8. 'Report of the Mineralogical Lecturer', in Queensland Dept of Mines, *Annual Report*, 1886, 88-93.
9. *Queenslander*, 16 June 1888, 946.
10. Queensland Dept of Mines, *Annual Report*, 1888, 111-13.
11. R. L. Jack, 'Report to the Minister for Mines', in *Queenslander*, 23 June 1888, 987.
12. *ADB*, vol. 4, 466; *Who's Who in Mining and Metallurgy* (London, 1910), 67.
13. *QldPD*, 19 October 1894, 983; Qld Royal Commission on Goldmining, *Report and Evidence* (Brisbane, 1897), 70; *Northern Miner*, 23 October 1891, 1.
14. Qld Dept of Mines, *Annual Report*, 1889, 125-6.
15. Gympie School of Arts and Mines Committee, *Minutes*, 2 April, 21 July 1890; Gympie School of Mines, *Minutes*, 15 January 1896, 25 January 1898; Gympie School of Mines, *Annual Report*, 1899, in Minute Book, OML, Brisbane, OMEM 1/1-1/3.
16. Qld Royal Commission, *Report and Evidence*, 237, 268.
17. *NSWPD*, 31 October 1893, 732-3; 29 August 1894, 87; 10 November 1896, 4872-6; 12 November 1896, 5074.
18. NSW Dept of Mines, *Annual Report*, 1887.
19. Vic. Dept of Mines, *Annual Report*, 1897-1900.
20. *VicPD*, 14 December 1898, 3741; Perry, *Ballarat*, 172.
21. WA Royal Commission on Mining, *Report and Evidence* (Perth, 1898), in *WAMVP*, 1898, Paper 20. For Parkes see p. 444, Callahan, p. 391.
22. Donald Clark, *West Australian Mining and Metallurgy* (Sydney, 1904), 86.
23. *WAPD*, 4 September 1900, 214.
24. *WA Yearbook*, 1902-04, 962.
25. See discussion in H. C. Hoover, 'The Future Gold Production of Western Australia', *TIMMet*, 1904-05, 14.
26. Clark, *West Australian*, 90.
27. *The Golden West* (Coolgardie, 1899), 12; *AMS*, 24 August 1899, 149.
28. It received not a single mention when the School of Mines Bill reached Parliament in 1894. See *QldPD*, 19 October 1894, 983-5; 13 November 1894, 1206-7.
29. Qld Royal Commission, *Report and Evidence*, 56, 78-9, 81, 114, 135, 151, 189.

30. *Ibid.*, 115, 162, LIV, 193-7.
31. *QldPD*, 26 October 1899, 637.
32. *AMS*, 30 July 1896, x; M. S. Churchward and P. Milner, *The Principal Engineering Establishments in Victoria in the Period 1842-1945*, Dept of Mechanical and Industrial Engineering, University of Melbourne, 1988, vol. IV, 95.
33. Churchward and Milner, *Engineering Establishments*, vol. IV, 23; *AMS*, 7 January 1897, 1579, 21 January 1897, 1601-2.
34. J. Smith, *The Cyclopedia of Victoria*, vol. 2 (Melbourne, 1904), 372-3; Churchward and Milner, *Engineering Establishments*, vol. II, 9-10.
35. Clark, *West Australian*, 77.
36. See advertisements in Chamber of Mines of WA, *Monthly Report*, e.g. 1904.
37. J. E. Clennell, 'Analytical Work in Connection with the Cyanide Process', *TIMMet*, 1902-03, 367; G. A. Byrn, 'The Laboratory in its Relation to the Cyanide Process', *TAusIMEng*, 1897, 173-85; N. Danvers-Power to Editor, *AMS*, 26 August 1897, 216.
38. J. S. Macarthur, 'Gold Extraction by Cyanide: A Retrospect', *JSCI*, 1905, 311-15.
39. Vic. Royal Commission, *Report and Evidence*, 765. For some other views, see pp. 856, 777-8, 193.
40. J. M. Maclaren, *TAusIMEng*, 1900-01, 379-400. See also advertisements, e.g. *Northern Miner*, 10 November 1900.
41. Vic. Royal Commission, *Report and Evidence*, 827; Ballarat School of Mines, *Annual Report*, 1892, in *Calendar for 1893*, 86-7.
42. Ballarat School of Mines, *Calendar for 1902-3*, 33-4, 46.
43. SA School of Mines, *Annual Report*, 1894, 48.
44. University of Melbourne, *Calendar for 1893*, 39.
45. Metallurgy I Paper, in University of Sydney, *Calendar*, 1893, 1894.
46. A. Liversidge, *JRSNSW*, 1894, 185-8.
47. A. R. Weigall, to Qld Royal Commission, *Report and Evidence*, Appendix P, 272-3.
48. *Calendar of the Sydney Technical College*, 1889-97.
49. Charters Towers School of Mines, *Syllabus for 1905*.
50. H. C. Jenkins, *TIMMet*, 1904-05, 188-265, esp. 199.
51. James Park, *The Cyanide Process* (London, 1904). George Trail's copy is owned by the author.
52. Dixon entry in *Who's Who in Mining and Metallurgy* (London, 1910).
53. *Ibid.*, Leggo entry; Ballarat School of Mines, *Calendar for 1902-3*; F. G. Brinsden, *TAusIMEng*, March 1910.
54. Clark, *West Australian*, 94.
55. *AMS*, 12 February 1903, 217.
56. Aarons entry, *Who's Who in Mining and Metallurgy*.
57. *Proceedings of the Australasian Institute of Mining Engineers Annual Meeting*, 1898-1903, 3-12.
58. Chamber of Mines of WA, *Monthly Report*, August 1903, 144-5.

59. Ballarat School of Mines, *Annual Report*, 1891, 83-97; *Calendar for 1902-03*, 46; Clark, *West Australian*, 51-2; Perry, *Ballarat*, 584.
60. For details see J. H. Todd, 'Transfer and Dependence: Aspects of Change in Australian Science and Technology', unpublished Ph.D. thesis, University of New South Wales, 1991, appendix 7.
61. 'Stawell School of Mines', *AMS*, 24 August 1899, 148-9.
62. *Who's Who in Mining and Metallurgy*, 25.
63. *Ibid.*, entry for Edgar.
64. *AMS*, 3 January 1901, 1-2.
65. Vic. Royal Commission, *Report and Evidence*, 824-31.
66. *Ibid.*, xvii; Ballarat School of Mines, *Annual Report*, 1890, 11-12.
67. SA Board of Technical Education, 'Report', 23.
68. *SA Register*, 10 June 1889, 7; SA School of Mines, *Annual Report*, 1889, 22.
69. SA School of Mines, *Annual Report*, 1896, 24; A. W. Norrie (ed.), *Australian Mining and Metallurgy* (Melbourne, 1953), 169-70.
70. NSW Dept of Mines, *Annual Report*, 1892, 14; 1894, 17-18; 1900, 67-9.
71. Vic. Royal Commission, *Report and Evidence*, 72-3.
72. SA School of Mines, *Annual Report*, 1896, 40.
73. *The Eagle*, 27 May, 23 June, 10 August, 5 October, 30 November 1895; *Charters Towers Mining Standard*, 20 August 1896; Qld Royal Commission, *Evidence*, 16, 70-8, 33-5, 123, 187.
74. WA Royal Commission on Mining, *Report and Minutes of Evidence* (Perth, 1898).
75. *QldPD*, 26 October 1899, 633-5.
76. *Ibid.*, 637.
77. W. Poole, in discussion of C. F. V. Jackson, *TAusIMEng*, August 1910, 220.
78. *WAPD*, 7 June 1900, 309, 4 July 1901; *Western Argus*, 26 February 1901, 35; John Laurent, 'Mechanics' Institutes and the Labor Movement: A Case Study of the West Australian Goldfields, 1895-1917', *Melbourne Studies in Education*, 1987-88, 81-98.
79. Qld Dept of Mines, *Annual Report*, 1901, 170; 1911, 188-91. Also Charters Towers School of Mines, *Syllabus for the Year 1905* (Brisbane, 1904), 4-6.
80. Kalgoorlie School of Mines, *Syllabus for 1911*, 23, 30, 70-4; K. S. Blaskett, 'Ore Treatment at Western Australian Gold Mines', *Chemical Engineering and Mining Review*, 10 November 1951, 48.
81. H. H. Dunkin and H. K. Worner, 'Mining and Metallurgical Education in Victoria', *Proceedings of the Australian Institute of Mining and Metallurgy*, 1950, 297-339; Norrie, *Australian Mining*, 168-73.
82. H. C. Jenkins, *TIMMet*, 1904-05, 188.
83. A. Jarman and E. le Gay Brereton, 'Laboratory Experiments on the use of Ammonia and its Compounds in Cyaniding Cupriferous Ores and Tailings', *TIMMet*, 1904-05, 289-334; E. le Gay Brereton, 'The Ammonia-Copper-Cyanide Process', *ibid.*, 1905-06, 433-44.

84. *AMS*, 29 October 1903, 578.
85. List of Members, *TAusIMEng*, 31 March 1912, 23-46.

13 TRANSFER, DIFFUSION AND LEARNING

1. See Jan Fagerberg, 'Why Growth Rates Differ', in G. Dosi, C. Freeman, R. Nelson, G. Silverberg, L. Soete (eds), *Technical Change and Economic Theory* (London, 1988), 432-57.
2. On changing perceptions, see Roy MacLeod, 'The Contradictions of Progress: Reflections on the History of Science and the Discourse of Development', *Prometheus*, 10 (1992), 260-84. On the shift in focus of studies of Third World technology, see Martin Fransman, 'Technological Capability in the Third World: An Overview and Introduction . . . ', in Martin Fransman and Kenneth King, *Technological Capability in the Third World* (London, 1984), 3-30.
3. Carlota Perez and Luc Soete, 'Catching Up in Technology: Entry Barriers and Windows of Opportunity', in Dosi *et al.*, *Technical Change*, 458-530; Roberto Camagni (ed.), *Innovation Networks: Spatial Perspectives* (London, 1991), 128.
4. Mark Dodgson, 'Strategy and Technological Learning', in Rod Coombs, Paolo Saviotti and Vivien Walsh (eds), *Technological Change and Company Strategies: Economic and Sociological Perspectives* (London, 1992), 136-63. On whether dicontinuities may be 'competence-enhancing' or 'competence-destroying', see Michael Tushman and Philip Anderson, 'Technological Discontinuities and Organization Environments', in Andrew Pettigrew (ed.), *The Management of Strategic Change* (Oxford, 1987), 89-127.
5. G. Dosi, 'The nature of the innovative process', in Dosi *et al.*, *Technical Change*, 221-30.
6. R. E. Evenson and H. P. Binswanger, 'Technology Transfer and Research Resource Allocation', in H. P. Binswanger and V. W. Ruttan (eds), *Induced Innovation: Technology, Institutions and Development* (Baltimore, 1978), 175-9.
7. Nathan Rosenberg, *Perspectives in Technology* (Cambridge, 1976), 173.
8. Nathan Rosenberg, *Inside the Black Box* (Cambridge, 1982), 245.
9. For an overview on the development of the diffusion literature, see Everett Rogers, *The Diffusion of Innovations*, 3rd edn (New York, 1983).
10. R. Coombs, P. Saviotti and V. Walsh, *Economics and Technical Change*, (London, 1987), 124-34; P. Stoneman and N. Ireland, 'The Role of Supply Factors in the Diffusion of New Process Technology', *Economic Journal*, March Supplement (1983), 65-77; J. S. Metcalfe, 'Impulse and Diffusion in the Study of Technical Change', *Futures*, 13 (1981), 347-59.
11. Paul Stoneman, *The Economic Analysis of Technological Change* (Oxford, 1983), 72.

12. Edwin Mansfield, 'Technical Change and the Rate of Imitation', *Econometrica*, 1961, 741-66; *Industrial Research and Technological Innovation* (New York, 1968), chapter 7.

13. Camagni (ed.), *Innovation Networks*; L. Gelsing, 'Innovation and the Development of Industrial Networks', in B. Lundvall (ed.), *National Systems of Innovation: Toward a Theory of Innovation and Interactive Learning* (London, 1992); Michael E. Porter, *The Competitive Advantage of Nations* (London, 1990).

14. J. W. McCarty, 'The Staple Approach in Australian Economic History', *Business Archives and History*, 4 (1964), 1-22.

15. Geoffrey Blainey, 'Technology in Australian History', *Business Archives and History*, 4 (1964), 117-37.

16. Rosenberg, *Black Box*. See Joel Mokyr, *The Lever of Riches: Technological Creativity and Economic Progress* (New York, 1990), for an opposing emphasis.

17. e.g. Mara Wilkins, 'The Role of Private Business in the International Diffusion of Technology', *Journal of Economic History*, 39, 166-88. See also Peter Heller, *Technology Transfer and Human Values: Concepts, Applications, Cases* (University Press of America, 1985).

18. Paul David, 'The Mechanisation of Reaping in the Ante-bellum Midwest', in H. Rosovsky (ed.) *Industrialization in Two Systems: Essays in Honour of Alexander Gerschenkron* (London, 1966), 3-28.

19. J. S. Metcalfe, 'The Diffusion of Innovation: An Interpretive Survey', in Dosi *et al.*, *Technical Change*, 560-89.

20. David Mowery and Nathan Rosenberg, *Technology and the Pursuit of Economic Growth* (Cambridge, 1989), 13.

21. Fagerberg, 'Growth Rates', 440.

22. e.g. Robert Solo, 'The Capacity to Assimilate an Advanced Technology', *American Economic Review: Papers and Proceedings*, May 1966, 91-7; Heller, *Technology Transfer*, the essays in Fransman and King, *Technological Capability*; in A. C. Smali (ed.) *Technology Transfer: Geographic, Cultural and Technical Dimensions* (Westport, 1985); and in Evenson and Ruttan, *Induced Innovation*.

23. For a policy-oriented study of the need for developing countries to build up their technological capability, see J. L. Enos, *The Creation of Technological Capability in Developing Countries* (London, 1991).

24. Kenneth Arrow, 'The Economic Implications of Learning by Doing', *Review of Economic Studies*, 29 (1962), 155-73.

25. Rosenberg, *Black Box*, 121-4, and *Perspectives*, 147, 166, 197-9.

26. Martin Bell, 'Learning and the Accumulation of Industrial Technological Capacity in Developing Countries', in Fransman and King, *Technological Capability*, 187-209.

27. C. Perez, 'New Technologies and Development', in C. Freeman and B. Lundvall (eds) *Small Countries Facing the Technological Revolution* (London, 1988), 89. In same volume see Bjorn Johnson, 'An Institutional Approach to the Small-Country Problem', 279-97.

28. H. R. Seddon, *Diseases of Domestic Animals in Australia*, vol. 1, part 5, (Canberra, 1955), 38.

29. Bengt-Ake Lundvall, 'Innovation as an Interactive Process: from User-producer Interaction to the System of Innovation', in Dosi *et al.*, *Technical Change*, 349-69.

30. A. K. Huntingdon, 'Presidential Address', *TIMMet*, 1893-94, 160-1.

31. Perez, 'New Technologies'.

32. Such as the Fink Commission on technical education in Victoria, and similar inquiries in other states. See Alan Barcan, *A History of Australian Education* (Melbourne, 1980).

14 COLONIAL SCIENCE: THE INTELLECTUAL BRIDGE

1. For a classification of papers to AAAS from 1888 to 1923, see Ian Inkster, 'Scientific Enterprise and the Colonial "Model": Observations on Australian Experience in Historical Context', *SSS*, 15 (1985), 694-7.

2. See, for instance, contributions in Nathan Reingold and Marc Rothenberg (eds), *Scientific Colonialism: A Cross-cultural Comparison* (Washington, 1987) and R. W. Home and Sally Kohlstedt (eds), *International Science and National Scientific Identity* (Dordrecht, 1991).

3. Stephen Hill, *The Tragedy of Technology* (London, 1988), chapter 8.

4. Thomas Hughes, 'The Seamless Web', in Brian Elliott (ed.), *Technology and Social Process* (Edinburgh, 1988), 9-19; Michel Callon, 'The Dynamics of Techno-Economic Networks', in R. Coombs, P Saviotti and V. Walsh (eds), *Technological Change and Company Strategies* (London, 1992), 72-102.

5. For instance, Nathan Rosenberg, *Technology and the Pursuit of Economic Growth* (New York, 1989).

6. Inkster, 'Scientific Enterprise'.

7. Edward A. Shils, 'Towards a Modern Intellectual Community', in James Coleman (ed.), *Education and Political Development* (Princeton, 1965), 498-500. See also *The Intellectuals and the Powers and Other Essays* (Chicago, 1971), esp. chapters 17-19.

8. Donald Clark, 'Notes on the Solubility of Gold-Silver Alloys in Cyanide of Potassium Solutions', *JRSVic*, 1898, 47-51.

9. A. Jarman and E. le Gay Brereton, 'Laboratory Experiments on the Use of Ammonia and its Compounds in Cyaniding Cupriferous Ores and its Tailings', *TIMMet* 1904-05, 289-334; E. le Gay Brereton, 'The Ammonia-Copper-Cyanide Process', *ibid.*, 1905-06, 433-44.

10. W. A. Dixon, 'Note on the So-called "Selective Action" of Cyanide of Potassium for Gold', *TIMMet*, 1897-98, 88-93; also published in *The Australian Technical Journal*, 28 March 1898, 54-60.

11. NSW Dept of Mines, *Annual Report*, 1897, appendices A and B, 22-3.

12. For references see relevant chapters.

13. C. Perez, 'New Technologies and Development', in C. Freeman and B. Lundvall (eds), *Small Countries Facing the Technological Revolution* (London,

1988), 92. See also C. Perez and L. Soete, 'Catching Up in Technology: Entry Barriers and Windows of Opportunity', in G. Dosi, C. Freeman, R. Nelson, G. Silverberg, L. Soete (eds), *Technical Change and Economic Theory* (London, 1988), 458-79.

14. George Bindon and David Miller, ' "Sweetness and Light'': Industrial Research in the Colonial Sugar Refining Company, 1855-1900', in R. W. Home (ed.), *Australian Science in the Making* (Melbourne, 1988), 170-94.

15. Robert Greig-Smith, 'Presidential Address', *JRSNSW*, 1916, 12.

16. A. G. Lowndes (ed.), *South Pacific Enterprise* (Sydney, 1956), 35-6, 41.

17. J. Cummins, *A History of Medical Administration in New South Wales* (Sydney, 1979), 69-81.

18. On Melbourne's public health crisis, see Gregory Truncheon, 'Germ Theory: Practical Implications for Medicine in Victoria', in K. Attwood and R. W. Home (eds), *Patients, Practitioners and Techniques* (Melbourne, 1985), 139-54. For a summary of Cherry's Board of Health work, see University of Melbourne, *Calendar*, various years, and 'Summary Report of Bacteriological Department' in papers submitted to the University of Melbourne Council, 1911/7, in the University of Melbourne Archives.

19. T. W. E. David, 'President's Address', *PLSNSW*, 1895, 137-43; Linnean Society of NSW, Papers of Sir William Macleay, 1874-88, ML MSS 2009116X.

20. Bruno Latour, *Science in Action* (Milton Keynes, 1987).

15 TOWARD AN AUSTRALIAN SYSTEM

1. Dieter Sengaas, *The European Experience* (New Hampshire, 1985), 60.

2. Thomas Hughes, 'Machines, Megamachines, and Systems', in Stephen Cutliffe and Robert Post (eds), *In Context: History and the History of Technology* (London, 1989), 106-19; Nathan Rosenberg (ed.), *The American System of Manufactures* (Edinburgh, 1969).

3. Bengt-Ake Lundvall, 'Innovation as an Interactive Process: From User-Producer Interaction to the National System of Innovation', in G. Dosi, C. Freeman, R. Nelson, G. Silverberg, L. Soete (eds), *Technical Change and Economic Theory* (London, 1988), chapter 17. See also R. Nelson (ed.), *National Innovation Systems* (New York, 1993).

4. A term coined by Thomas Hughes to take account of the specific influences of the new environment in which a transferred technology becomes embedded. See Thomas Hughes, 'The Evolution of Large Technological Systems', in W. Bijker, T. Hughes and T. Pinch (eds), *The Social Construction of Technological Systems* (London, 1987), 68-70.

5. Qld Royal Commission on Goldmining, *Report and Minutes of Evidence* (Brisbane, 1897), 145.

6. Memo from NSW Chief Inspector of Stock, 5 June 1917, in ASF 109, AONSW 12/3522.

7. Roy MacLeod, 'Scientific Advice for British Imperial India', *Modern Asian Studies*, 1975, 434-84; 'The Alkali Acts Administration, 1863-84', *Victorian*

Studies, 1965, 85-112; and 'Statesmen Undisguised', *American Historical Review*, 1983, 1386-405.

8. T. G. Parsons, 'Technological Change in the Melbourne Flour-Milling and Brewing Industries, 1870-90', *Australian Economic History Review*, 1971, 133-41; E. M. Sigsworth, 'Science and the Brewing Industry, 1850-1900', *Economic History Review*, 1965, 536-50; J. C. MacCartie, 'Some Brewing— and Other—Reminiscences', *The Australian Cordial Maker*, 24 January 1908, 26-7.

9. A. G. Lowndes (ed.), *South Pacific Enterprise* (Sydney, 1956), chapter 2; George Bindon and David Miller, 'Sweetness and Light', in R. W. Home (ed.), *Australian Science in the Making* (Melbourne, 1988), 170-94.

10. Australian Academy of Technological Sciences and Engineering, *Technology in Australia, 1788-1988* (Melbourne, 1988), 101-2.

11. M. A. O'Callaghan, 'Dairy Bacteriology', *Agricultural Gazette of NSW*, April 1899, 291.

12. 'Report of the Dairy Expert', in NSW Dept of Agriculture, *Annual Report*, 1901; 1902.

13. K. T. H. Farrer, *A Settlement Amply Provided* (Melbourne, 1980), chapter 11; Parsons, 'Technological Change'.

14. A. R. Callaghan and A. J. Millington, *The Wheat Industry in Australia* (Sydney, 1956); Aust. Academy of Technological Sciences, *Technology*, 634-6; Ian Rae, 'Chemists at ANZAAS', in Roy MacLeod (ed.), *Commonwealth of Science* (Melbourne, 1988), 168; Bruce Davidson, 'Developing Nature's Treasures', *ibid.*, 274.

15. Don Fraser, 'Bridges', in Don Fraser (ed.), *Sydney: From Settlement to City* (Sydney, 1989), 99-124.

16. Ken Slater, 'Acid to Ashes: 19th Century Electrical Technology and the Foundations of Electricity Supply in Melbourne', unpublished M.A. thesis, La Trobe University, 1981.

17. Ann Moyal, *Clear Across Australia: A History of Telecommunications* (Melbourne, 1984), 108-13.

18. Australian Academy of Technological Sciences, *Technology*, 863-4; 920-1.

19. *Ibid.*, 871-3, 909-10.

20. *Ibid.*, 876; Paul Savage, *With Enthusiasm Burning* (Sydney, 1974).

21. Australia's share of British exports fell from 8.5 per cent in the 1880s to 5.5 per cent by 1905-09, while the British component in Australia's imports fell from 70 per cent to 61 per cent. At the same time Britain's share of Australia's exports fell from 75 per cent to 47 per cent, as Australia became far less dependent on Britain as a market. See Barrie Dyster and David Meredith, *Australia in the International Economy* (Melbourne, 1990), 51-3.

22. Ian Inkster, 'Intellectual Dependency and the Sources of Invention', *History of Technology*, 1990, 59.

23. Quoted in G. J. R. Linge, *Industrial Awakening: A Geography of Australian Manufacturing, 1788-1890* (Canberra, 1979), 219.

24. *Ibid.*, 433.

25. 'Jacques with Compliments', unpublished typescript in NBAC/ANU, Jacques Bros, 125/26.
26. See, for instance, J. W. McCarty, 'Australia as a Region of Recent Settlement in the Nineteenth Century', *Australian Economic History Review*, 1973, 148-67.
27. Dyster and Meredith, *Australia*, 16.
28. Gerschenkron is well known for his thesis that the later the start of industrialisation, the greater the role of the state: Alexander Gerschenkron, *Economic Backwardness in Historical Perspective* (Cambridge, 1966).
29. On 3 June 1992, the High Court of Australia handed down a landmark decision on what has become known as 'the Mabo case': it threw out the notion of *terra nullius*, and thereby recognised that Aboriginal people were dispossessed of their land, correcting the widespread misconception that Australia did not belong to anyone before it was settled by the British in 1788.
30. e.g. Sidney Pollard, *Peaceful Conquest: The Industrialization of Europe, 1760-1970* (Oxford, 1981); Michael E. Porter, *The Competitive Advantage of Nations* (London, 1990); Christopher Freeman and Bengt-Ake Lundvall (eds), *Small Countries Facing the Technological Revolution* (London, 1988).
31. Linge, *Industrial Awakening*, 8.
32. Slater, 'Acid', 230-2.
33. Linge, *Industrial Awakening*. This point emerges in several places but is highlighted in the introductory chapter.
34. See Pollard, *Peaceful Conquest*, for instance, on Eastern Europe.
35. Barrie Dyster, 'Argentine and Australian Development Compared', *Past and Present*, 1979, 96.

Index